Experiencing Science

Experiencing Science

Jeremy Bernstein

A Dutton Paperback
E. P. Dutton • New York

This paperback edition published in 1980 by E. P. Dutton, A Division of Elsevier-Dutton Publishing Co., Inc., New York

For information contact:
E.P. Dutton, 2 Park Avenue, New York, N.Y. 10016

Library of Congress Cataloging in Publication Data
Bernstein, Jeremy, 1929—
Experiencing science.
Bibliography: p. 266
Includes index.
CONTENTS: Two faces of physics: Kepler, harmony of the world. Rabi, the modern age.—Three faces of biology: Lysenko, enemies of the people. A sorrow and a pity, Rosalind Franklin and The double helix. Lewis Thomas, life of a biology watcher.—Fact and Fantasy: Extrapolators, Arthur C. Clarke. Clarke. Calculators, self-replications. Innovators, Gödel's theorem.
1. Scientists—Biography. 2. Science—History.
I. Title.
Q141.B395 509'.2'2 77-20415

ISBN: 0-525-47636-9

Published simultaneously in Canada by
Clarke, Irwin & Company Limited, Toronto and Vancouver

10 9 8 7 6 5 4 3 2 1

CONTENTS

INTRODUCTION

THE WRITING collected in this book is work that I have done over the last decade. Much of it has appeared in *The New Yorker;* one of the pieces appeared in *The New York Times Magazine* under the flamboyant title—not my invention—"When the Computer Procreates"; and some of it has never appeared anywhere. All of this writing deals with the scientific experience in all of its manifestations, including science fiction and a new kind of fiction that I have recently been experimenting with. Over the last twenty-five years, I have been a "science watcher" in something of the same sense that Lewis Thomas, if he will forgive the comparison, has been a "biology watcher." My own special field of science is the theoretical physics of elementary particles. But while working away on particular problems I have lifted my head, like a sort of grazing giraffe, to look around at the rest of the trees in the forest. I am not an expert in everything, and a great deal of this writing has involved long, and often rather painful, processes of self-education. After my *New York Times Magazine* piece on computers appeared—to strong and somewhat mixed reactions—a very respected colleague asked me if I really understood what I was talking about. One's scientific colleagues are not shy about asking questions like that. In response I would like to repeat a story, as told by Aage Petersen, one of Niels Bohr's assistants, a story that appears to have been one of Bohr's favorites. Petersen's version is as follows: In an isolated village there was a small Jewish community. A famous rabbi once came to the neighboring city to speak, and because the people of the village were eager to learn what the great teacher would say, they sent a young man to listen. When he returned he said, "The rabbi spoke three times. The first talk was brilliant, clear and simple. I understood every word. The second was even better; deep and subtle. I didn't understand much but the rabbi understood all of it. The third was by far the finest; a great and unforgettable experience. I understood nothing and the rabbi himself didn't understand much either." Over the years I have felt at times like the student and at times like the rabbi. However, even when my own understanding was not as deep as I wished it to be, I have felt that it was worthwhile to try to communicate to "the people of the village" what I thought I did understand and to leave it to my successors to do better. In this day and age the experience of science is something that affects all of us, and we scientists have a crucial obligation to describe it as best we can.

The reader may be interested in knowing something of how all this came about. I entered Harvard in 1947, at the age of seventeen, with no clear idea of what I intended to do with my life. I knew, or thought I knew, that the one thing I did well was to write and so I felt that I might become a journalist. Becoming a scientist was the last thing I had in mind. I was a near cipher in high school science and, although good at various kinds of mathematics, I always regarded such things as "subjects" rather than disciplines that could arouse the passionate concerns of serious people. Nonetheless I had a random, and rather romantic, curiosity about Einstein and his theory of relativity, which at the time I was convinced was only understood by five people in the world. (When Sir Arthur Eddington was asked, in the 1920s, if it were true that only three people in the world understood the theory of relativity, he replied, "Who is the third?") I was therefore quite surprised when I discovered that at least the special theory of relativity was a standard subject even in the courses that Harvard had then designed for non-scientists like myself. I took several of these, including some given by the late Philipp Frank, the noted biographer of Einstein who had succeeded him after Einstein's brief tenure in Prague. I also got my first introduction to quantum physics from Professor Frank and became, thanks largely to him, hooked on modern physics. It took me years of work, however, to fill in the terrible gaps in my own scientific and mathematical education, much of which I did by self-study. (I was very amused recently when I was asked to teach an advanced course in classical electricity and magnetism, a fundamental subject for any aspiring physicist, to recall that in all of the eight years I spent at Harvard as a student I had never taken a single course, on any level, in this subject.) In any case, all thoughts of writing—at least in a professional sense—were suppressed during this time and indeed until I reached the age of thirty.

Nonetheless, somewhere deep inside I felt that I was a "writer." (In one of Antonioni's films a woman remarks that she has all the vices, except that she does not "choose to practice any of them.") I kept some notebooks with bits and pieces of writing in them and wrote incredibly long letters to people, one of which I sent to *The New Yorker.* This "letter," written in 1960, concerned the teaching of physics in a summer school on the island of Corsica and a girl named Annie—a native of the island—with whom many of us were in love. I sent it off—unsolicited of course—from the Brookhaven National Laboratory on Long Island, where I was working full-time, and after many months received word from the magazine that they would like to buy it. "Word" in this case, came directly, by phone, from William Shawn, the editor of the magazine. What Mr. Shawn said, as I recall, was, "We liked your piece on Corsica and we wonder if it would be all right with you if we published it." This is what I remember him saying, although I was in such a daze at the time that I am not absolutely sure. In any event, he invited me to come to New York to see him, an invitation I accepted at once.

Our meeting lasted, as I recall, well over an hour. At the time I did not find anything unusual in this, but I have come to realize over the years that this was, for him, an expensive investment of time. I think he wanted to see what sort of creature had emerged from the laboratory to produce a piece of writing suitable for the magazine and to see, I imagine, if that was it or whether I might show some promise of producing something else. Our conversation ranged over everything that makes up the life of a young scientist, and when it was over he said to me that there was a kind of writing I could do for the magazine that no one up to that time had tried to do for them. I could write about science as a form of experience—my own experience and the experience of people I knew—and this is what I have been trying to do ever since.

Most of the writing in this book falls into standard categories. It is what *The New Yorker* editors would describe as "fact"—reportage and commentary about the subject at hand. There are profiles of I. I. Rabi, Arthur Clarke, and Lewis Thomas, as well as essays about Kepler and the discovery of the double-helix structure of DNA, and there is a piece on self-replicating automata. I believe that the reader will find that, taken together, these writings, done at various times and places and with different aims and goals in mind, have a certain relatedness and internal consistency. In rereading them and bringing them up to date for the book I was surprised by this myself. I found that several of the articles seemed to be talking to each other, although the ostensible subject matters were rather different and the times at which they were written differed by several years. In trying to figure this out, I came to the conclusion that each time I wrote about something that really interested me, what I had learned about it made me want to learn more about it; therefore, when I wrote a new piece on something related, I put my new knowledge in it. It was as if the opus as a whole was continuing to grow organically. I hope the reader will feel as I do and that this is not simply a collection of random and unrelated articles. Although the theme of the collection is the large scientific experience, it is, after all, being filtered through the perceptions and developing experience of one individual. However, I have also included a somewhat strange fictional love story, which involves Gödel's theorem; the famous theorem, proved by Kurt Gödel in 1931, implics that no formal proof of the absolute consistency of mathematics is possible. The fundamental question of whether or not mathematics is consistent cannot be decided. I have been studying Gödel's proof, and its consequences, on and off for years. To this day I do not understand all of the details, and my story may explain why. So, in this matter, I am somewhere between the second and the third appearances of Bohr's rabbi. But while engaged in my last foray into the theorem I was struck by the question of why I didn't understand all of the details. What was getting in the way? What else was going on in my head, and in my life, while I was trying to understand the proof of the theorem? The more I thought about this the funnier it seemed to me—real life intruding on Gödel's theorem and vice versa. In due course I decided to write a story about *that* as well as about the theorem itself. Here was an experience of science which I think can be set down only in this way. Whether one wants to call the result fiction or something else does not seem very important to me. It is a different way of trying to write about science, and I am intrigued with the possibilities it opens up. Once, in discussing fiction with Mr. Shawn, he said that it was essential, in his view, for the writer to hold back nothing out of his own experience. If one is a scientist writing fiction, this fact should be evident in the work in some way. From my story about Gödel's theorem, whatever may be wrong with it, it is clear, I think, that it could only be written by a scientist.

From the preceding remarks, it is obvious that I am indebted to both Philipp Frank and William Shawn in similar ways. Each of them guided and inspired me at times of my life when I was taking my first uncertain steps in very new directions. Philipp Frank died some years ago, so I can only express my gratitude to his memory. To Mr. Shawn I simply express my fondest appreciation. In addition, I would like to acknowledge the editorial help of Gardner Bostford and Pat Crow, who edited *The New Yorker* profiles that appear here. The work is vastly better for their efforts. I am also pleased to thank Karen Kennerly, who helped me prepare the manuscript of this book.

JEREMY BERNSTEIN

New York City, February 7, 1978

ACKNOWLEDGMENTS

Permission to reprint material from the following sources is gratefully acknowledged.

Chapters 2, 3, 5, and 6 originally appeared in *The New Yorker*. Copyright © 1969, 1970, 1975, and 1978 by The New Yorker, Inc.

Chapter 7 originally appeared in *The New York Times*, February 15, 1976. Copyright © 1976 by The New York Times Company.

Excerpts from: Lewis Thomas, *Lives of a Cell: Notes of a Biology Watcher* (New York: The Viking Press, 1974). Copyright © 1974 by Lewis Thomas. Reprinted by permission of The Viking Press.

Excerpts from: Arthur Koestler, *The Sleepwalkers* (New York: The Universal Library, Grosset & Dunlap, 1963). Copyright © 1963 by Arthur Koestler. Reprinted by permission.

Excerpts from: Angus Armitage, *John Kepler* (New York: Roy Publishers, 1967). Copyright © 1967 by Angus Armitage. Reprinted by permission.

Excerpts from: Zhores A. Medvedev, *The Rise and Fall of T. D. Lysenko,* trans. I. Michael Lerner (New York: Columbia University Press, 1969). English translation Copyright © 1969 by Columbia University Press. Reprinted by permission of Columbia University Press.

PART I

Two Faces of Physics

IT IS HARD to imagine, at first sight, two physicists whose lives and careers have less in common than I. I. Rabi and Johannes Kepler. Kepler was a late sixteenth- and early seventeenth-century Protestant mystic; Rabi, profoundly Jewish, and profoundly rooted in the life of the twentieth century. Yet, it is the fact that both of these men, despite the inclinations they might have had to isolate themselves in their scientific studies, were forced by events to become part of the political and social world around them that brings them together at a common point. In both their lives the academy was never very far away from thc battlefield. In both their lives the sound of religious prejudice was never very far away from the window. Of course, in three centuries technique and technology have changed. Kepler cast horoscopes to predict the outcome of his emperor patrons' battle plans, while Rabi built radar for the United States Navy and helped Robert Oppenheimer make the atomic bomb. Kepler succeeded in persuading the elders of Leonberg not to burn his mother at the stake as a witch, while Rabi, despite all his efforts, did not succeed in saving Oppenheimer from a McCarthy-inspired witch-hunt. Both men lived through deeply troubled times and both men shared the feeling, beautifully expressed by Kepler, that "in vain does the God of War growl, snarl, roar and try to interrupt with bombards, trumpets and his whole tarantaran. . . . Let us despise the barbaric neighings which echo through these noble lands and awaken our understanding and longing for the harmonies."

1

KEPLER: HARMONY OF THE WORLD

IN SCIENCE as in the arts, sound aesthetic judgments are usually arrived at only in retrospect. A really new art form or scientific idea is almost certain at first to appear ugly. The obviously beautiful, in both science and the arts, is more often than not an extension of the familiar. It is sometimes only with the passage of time that a really new idea begins to seem beautiful. As has been said, "Pioneers occupy new land. Only later, one comes to understand that the cabins they built were really cathedrals." Even the creators of new scientific ideas often find their offspring unbearable. The history of science is full of great scientists who have rejected the most original parts of their own ideas or who have failed to esteem what was of greatest significance in their own work because the novelty of it made it impossible to place the ideas in any familiar context. Einstein's rejection of the quantum theory, in the discovery of which he had had a major part, is only thc most recent example.

Because of the fragmentation of contemporary culture, at least in the West, the extra-scientific influences on the acceptance or rejection of scientific theories are probably less important now than at any time in our intellectual history. But one should not forget that the Nazis forbade the acceptance of Einstein's theory of relativity, except as a computational trick, because it was "Jewish physics," just as one should not forget that until 1965 anyone in the state of Arkansas who taught Darwinian evolution was subject to prosecution. (The distinction the Nazis made between the theory of relativity as a "true" physical theory as opposed to its being a computational device which had to be used to obtain agreement with certain experimental results, to "save appearances," has, as we shall see, roots that stretch back to the beginnings of science.) And in modern China, where every effort has been made to unify culture within the confines of Maoist philosophy, there is an attempt to divine—in certain ideas of contemporary science and, most notably, elementary particle physics—illustrations of the maxims of Chairman Mao. To a Western reader, the maxims seem so vague and so general that it is hard to imagine how any genuine scientific idea could contradict them. If this were not the case, then Chinese science would suffer the same sort of paralysis and schizophrenia that Western science underwent during the late Middle Ages and the early Renaissance. No philosopher, and no scientist, either, can anticipate the course of scientific discovery. Discoveries can be fitted into a preconceived philosophical structure only if the structure is so loose that it can accommodate anything or if, as was the case with the Catholic Church during the rise of the New Astronomy in the sixteenth and seventeenth centuries, the structure is rigidly divided into segments whose propositions are judged to be outside the competence of scientific investigation and segments in which scientific ideas are allowed free reign. Such compartmentalization of ideas leads to schizophrenia, and if a schizophrenic has the polemic personality of a Galileo, the whole edifice may be shaken to the point where it cracks apart. The history of scientific ideas shows again and again that an ideological house divided against itself cannot stand intact for long.

The rise of the New Astronomy, whose principal architects

were Copernicus, Tycho Brahe, Kepler, Galileo, and Newton, may be dated from 1543, the year Copernicus published his *Book of the Revolutions*. (It was also the year he died; in fact, it was only on his deathbed that he was presented with a copy of his book.) It is a development that has fascinated historians. In the hundred years following the death of Copernicus—Newton was born in 1642 and published his *Principia* in 1687—modern science came into being. It was not an easy birth, and during the course of it an entire concept of the world was first destroyed and then reconstructed in a form beyond recognition. And the personalities involved were remarkable. The geniuses of the New Astronomy were almost the last scientists to work in an epoch in which scientific discovery was reasonably well integrated into the rest of intellectual life. They were the last of the natural philosophers (after them came the specialists), and the science they evolved has become our common-sense science—a science built on discoveries made with equipment that was scarcely more sophisticated than the unaided senses and that dealt with things within the range of ordinary human experience; for example, the motions of the planets. Since three centuries have now passed, one may separate out the enduring contributions from the chaff.

Among this group of geniuses, Johannes Kepler, who was born in the southwestern German town of Weil-der-Stadt on the twenty-seventh of December, 1571, has been of special interest to scholars. In addition to being one of the most original thinkers in the history of science, he had an extraordinarily open and appealing personality. Unlike Copernicus and Newton, who were extremely secretive, and unlike Galileo, who was not eager to share ideas (he announced some of his most important telescopic observations in the form of coded anagrams so that he would not have to announce discoveries publicly until he could confirm them), Kepler wrote letters, books, pamphlets in reams about his discoveries and his life. He was never content just to announce his discoveries in final, polished form; he shared the process of discovery by taking his readers with him every hesitant step of the way, discoursing all the while on his own fallibility. Imagine a modern scientific tract containing comments like "Why should I mince my words? The truth of Nature, which I had rejected and chased away, returned by stealth through the back door, disguising itself to be accepted. . . . I thought and searched, until I went nearly mad, for a reason why the planet preferred an elliptical orbit. . . . Ah, what a foolish bird I have been!" So commenting, Kepler reveals in his major work, *Astronomia Nova,* perhaps his greatest discovery—that the planet Mars and all the planets move in elliptical, not circular, orbits around the sun. Among the more popular studies of Kepler and the New Astronomy, three of the most interesting are *The Sleepwalkers,* by Arthur Koestler, *Kepler's Dream,* edited by John Lear, and *John Kepler,* by the British historian of science Angus Armitage. Such is Armitage's dedication to popularization that he even translated Kepler's first name into English. These books contribute in complementary ways to the study of Kepler and his epoch. Koestler writes with the touch of a novelist, and in addition to brilliantly illuminating the personalities involved he uses them as illustrations for his conception of the role of the unconscious in the process of invention. (Thus the title of his book.) Kepler, like most great scientists, had a remarkable subliminal instinct for scientific truth—a "pipeline to God," as the phrase goes. However—and this is equally astonishing, and also characteristic of some of the greatest scientists, including Einstein—he was often driven by instincts of harmony and beauty that in some instances appear to have been totally misguided. Koestler's book is sketchy about the period between Kepler and Newton, but Armitage gives a clear outline of how Kepler's ideas came to the attention of Newton and in what form he assimilated them. *The Dream,* a translation of Kepler's *Somnium, Sive Astronomia Lunaris* (the translation, by Patricia Frueh Kirkwood, to which Lear has appended a fascinating explanatory introduction, is the first in English), is something else again. It is a work of science fiction. Indeed, many people consider it the first great work of science fiction, and, as is true of the other, and rare, great works of science fiction, there is much more to it than at first meets the eye.

Much of Kepler's life was filled with hardship. At the age of twenty-six he produced a kind of genealogical horoscope in which he presented thumbnail sketches of his family, including himself. They are uniformly grim and disturbing:

> Henrich, my father, born 1547, 19 January. . . . A man vicious, inflexible, quarrelsome and doomed to a bad end. Venus and Mars increased his malice. [This is a horoscopic reference. Throughout his life as a professional astronomer, under various royal patronages, Kepler was obliged to design horoscopes. Apart from the fact that this helped him eke out a living, he felt, in the spirit of the age in which he lived, that the planets and stars influenced the fate of men. As he grew older, his view of astrology became more and more abstract, and in the end he seemed to feel that planetary conjunctions provided simply the general terrestrial atmosphere within which men worked out their own destinies. As a court astrologer he was sometimes in a position to influence political events, but, often to the disappointment of his royal patrons, his intellectual honesty was such that he was unwilling to take advantage of these opportunities.] Jupiter combust in descencion made him a pauper but gave him a rich wife. Saturn in VII made him study gunnery; many enemies, a quarrelsome marriage . . . a vain love of honors, and vain hopes about them; a wanderer. . . . 1577: he ran the risk of hanging. He sold his house and started a tavern. 1578: a hard jar of gun powder burst and lacerated my father's face. . . . 1589: treated my mother extremely ill, went finally into exile and died.

Kepler's father was a military mercenary, and he vanished from sight during a military expedition while Kepler was still in his teens. Kepler's mother, Katherina, whom, late in his career, he was barely able to save from being burned at the stake as a witch, hardly comes off much better in *her* horoscope: "Small, thin, swarthy, gossiping and quarrelsome, of a bad disposition." Koestler remarks, "The notes on his own childhood and youth, in the family horoscope, read like the diary of Job." Kepler, in his own horoscope, reports an endless chronicle of childhood misery and illness:

> On the birth of Johannes Kepler. I have investigated the matter of my conception, which took place in the year 1571, May 16, at 4:37 A.M. . . . My weakness at birth removes the suspicion that my mother was already pregnant at the marriage, which was the 15th of May. . . . Thus I was born premature, at thirty-two weeks, after 224 days, ten hours. . . . 1575 I almost died of smallpox, was in very ill health, and my hands were badly crippled. . . . 1577 On my birthday I lost a tooth, breaking it off with a string which I pulled with my hands. . . . 1585-86 During these two years, I suffered continually from skin ailments, often severe sores. . . . On the middle finger of my right hand I had a worm, on the left a huge sore. . . . 1591 The cold brought on prolonged mange. . . . 1592 I went down to Weil and lost a quarter florin at gambling. . . . At Cupinga's I was offered union with a virgin; on New Year's Eve I achieved this with the greatest possible difficulty.

And on and on.

The Keplers were part of a Lutheran enclave in a largely Catholic community, but, Armitage remarks:

> Following the Reformation, Latin schools had been founded up and down Württemberg under ducal patronage, to take the place of the former monastic schools. They were designed to educate promising lads for the Church and the civil service. As a boy of seven, Kepler, already marked out by precocious talent, entered the Latin school at Leonberg, where the family was then living.

Kepler distinguished himself in Latin studies, and in the autumn of 1589 he entered the University of Tübingen, the highest institute of theological study in the dukedom of Württemberg. His torments followed him to Tübingen, and, as a burgeoning prodigy, he was persecuted by his schoolmates. Later, he wrote a bitter portrait of himself:

> That man has in every way a doglike nature. His appearance is that of a little lapdog. His body is agile, wiry, and well-proportioned. Even his appetites were alike: he liked gnawing bones and dry crusts of bread, and was so greedy that whatever his eyes chanced on he grabbed; yet, like a dog, he drinks little and is content with the simplest food. His habits were similar. He continually sought the good will of others, was dependent on others for everything, ministered to their wishes, never got angry when they reproved him, and was anxious to get back into their

favor. He was constantly on the move, ferreting among the sciences, politics, and private affairs, including the lowest kind; always following someone else, and imitating his thoughts and actions. He is bored with conversation, but greets visitors just like a little dog; yet when the least thing is snatched away from him, he flares up and growls.

Such is the paradox of the human personality that, despite (or perhaps because of) a distraught and broken childhood, Kepler became one of the most determined seekers after cosmic harmony in the history of human thought. The transformation began at Tübingen. The discipline of a young seminarian required that Kepler attend the lectures of the astronomer Michael Mästlin, from whom he learned the workings of Copernican astronomy. Armitage notes:

> Writing a few years later, Kepler relates how "when I was at Tübingen six years ago, studying under the renowned master Michael Mästlin, I was impressed by the manifold inconvenience of the customary opinion concerning the Universe; and I took such delight in Copernicus, of whom Mästlin, in his lectures, used to make frequent mention, that I . . . used often to defend the doctrines of Copernicus in the candidates' disputations on physics."

To understand the full significance of Kepler's statement and of his subsequent work we must retrogress a bit into the history of astronomy. The ancient Greeks made profound advances in planetary astronomy. Aristarchus of Samos, who is thought to have been born in 310 B.C., had even conceived a sun-centered cosmology, in which the planets moved around a stationary sun, but with the decline of Greek civilization the heliocentric system was forgotten until Copernicus resurrected it, nearly 2,000 years later. It has been suggested that only the dross from Greek civilization managed to stay afloat through the Middle Ages and that the gold sank from view. Be that as it may, Europe in the late fifteenth and early sixteenth centuries had inherited some of the most naive aspects of Greek astronomy. The Greek astronomy that was generally accepted at the time of Copernicus derived mainly from Aristotle, Plato, and Ptolemy. Aristotle's physics was entirely qualitative and full of irrelevant biological metaphor. One of his qualitative laws, which had a pernicious effect on the subsequent development of astronomy, was his doctrine of "inertia," which came to be summarized in the Latin maxim *Omne quod movetur ab alio movetur,* or "Whatever is moved must be moved by another." (The correct version of the law of inertia, discovered by Galileo and then enunciated by Newton, in his ultimate synthesis of the New Astronomy, states that once a body is set in motion it will continue to move indefinitely in a straight line providing that no force acts on it.) Aristotle and Plato further reasoned that the earth was stationary and at the center of the universe. Plato had contributed the idea that celestial matter must move in uniform circular motion; that is, that the planets and the sun must move in circles around the stationary earth. To accord with the conception of inertia, the assumption was later made that the moving planets were attached to rigid crystalline spheres, which were thought to be maintained in their motion by angels.

The planets and the stars are often still pictured as being on the surface of the distant celestial sphere, the vault of the heavens. If the philosophers had had their way, the planets would have traced out great circles on this sphere as they moved at a uniform rate among the fixed stars. But it has long been known, through observation, that planetary motion is far from being uniform and circular. In particular, the planets appear to move more rapidly when they are close to the sun, and periodically they appear to reverse the direction of their orbital motion—the phenomenon of retrograde motion—and to travel in a loop before continuing on their original course. However, Plato and his followers insisted that these variations were merely a matter of "appearance." To preserve the concept of uniform, circular motion, the idea of the planetary epicycle was evolved, and it was developed in detail by the Hellenic school of astronomy, which culminated in the work of Ptolemy, who flourished in Alexandria in the second century A.D. An epicycle is a motion in the form of a circle that is centered on the perimeter of a larger circle—as if a planet were attached to the rim of a small Ferris wheel that, in turn, revolved about a point on the rim of a large Ferris wheel. But the theory of the epicycles, as they were worked out in detail in Ptolemy's *Almagest,* could not account for the observed planetary motions if the earth were assumed to be at the exact center of all of the Ferris wheels, and in Ptolemy's scheme the earth was placed at rest off-center within these large circles and the planets revolved in their epicycles, and often in epicycles on top of epicycles, around the center of the circles.

The mind boggles at the complexity of it. Koestler quotes Alfonso X of Castile, a medieval royal patron of astronomy: "If the Lord Almighty had consulted me before embarking upon the Creation, I should have recommended something simpler." Indeed, Ptolemy could not reconcile even the crude, naked-eye measurements of the planetary orbits that were known to him with the assumption that the planets moved at uniform speed on their various circles, and in the end he assumed that there might be a point within the big circles—the "equant" point—from which the motion would appear to be uniform if the earth were situated at that point, which it wasn't. It was this introduction of the equant by Ptolemy, which appeared to violate Plato's philosophical dictum that all heavenly bodies move uniformly in circles, that led Copernicus to revive the heliocentric-world theory. Both Armitage and Koestler make it clear that there were intimations of the heliocentric system in the work of other astronomers just before Copernicus began his work. But Copernicus amplified and crystallized their ideas. In his *Commentariolus* he said, "Having become aware of these defects [in the Ptolemaic system], I often considered whether there could perhaps be found a more reasonable arrangement of circles . . . in which everything would move uniformly about its proper center, as the rule of absolute motion requires." Copernicus's "arrangement of circles" involved moving the sun to the center of the universe and setting the earth in motion on its own circular orbit. Even this drastic step did not enable Copernicus to dispense with the epicycle theory; as Koestler points out, in the final version of his system he was forced to employ even more epicycles than Ptolemy had, to jibe with some new observations he had made of the planetary orbits. It was, in a sense, fortunate for the future of the heliocentric system that Copernicus was not a better observer—he made only sixty or seventy orbital sightings in his lifetime—and that his standards of accuracy were low; otherwise, he would have realized, as Kepler found out later, that his scheme, too, was a most inexact approximation of the actual planetary orbits. If his scientific standards had been higher, he might have abandoned the heliocentric system then and there.

As it was, Copernicus was extremely reluctant to publish his work, and it is unlikely that the *Book of the Revolutions* would have ever found its way to the printer if, late in Copernicus's life, a disciple, Georg Joachim von Lauchen, had not attached himself to the old man and literally snatched the manuscript away from him. It was not religious opposition that kept Copernicus from publishing. In fact, influential members of the Catholic Church knew about his work and encouraged him to publish, and not until 1616, at the time of Galileo, seventy-three years after Copernicus had died, was his book put on the Index. His hesitation reflected the fact that his system contradicted both Aristotelian physics and common sense, for everyday experience appears to indicate that the earth does *not* move, there being no sensation of motion. Apart from the simple fact that there is no obvious evidence of the earth's motion, there were other arguments against it, some of which appear to us as nonsense and some of which even a modern scientist can understand and accept. The nonsense arguments were of the order that if the earth moved, free-floating objects, such as birds, would be left behind, and these arguments were based on a faulty understanding of the law of inertia. One is aware of motion during acceleration from, say, a dead stop, but once a constant velocity has been attained—in, for instance, a train whose windows are curtained—there is no sensation of motion, and if someone in the train tosses a ball in the air, it will move with the train, since it shares the train's momentum. A *valid* argument held, however, that if the earth indeed was in motion, one should be able to detect that motion by watching the fixed stars, for were the earth to move, the position of the near stars should appear to change in relation to the position of the faraway stars, precisely as a nearby object viewed from a moving train appears to change position in relation to the position of faraway objects—a phenomenon known as "parallax." It was argued, and correctly, that the planetary model devised by Copernicus implied that stellar parallax should be visible from the earth, but at the time it was not. Nevertheless, men like Kepler, who insisted that Copernicus was right, said that the reason the changes of position had not been observed was that the stars were much farther away from the earth than had been imagined. To accept the Copernican theory therefore meant accepting the concept of a universe of enormous size—a thought that terrified the medieval mind. The phenomenon of stellar parallax was not actually observed until the nineteenth century, by the German astronomer Friedrich Wilhelm Bessel. His discovery established that stars *are* tremendously far away, but this does not seem to trouble the modern mind.

Book of the Revolutions aroused little or no interest at first, but half a century later, when Kepler was beginning his studies at Tübingen, enough of its contents had diffused into the general intellectual atmosphere that it was taught and debated by teachers as enlightened as Mästlin, Kepler's mentor. Upon graduating from the Faculty of Arts, Kepler matriculated at the Theological Faculty at Tübingen, but before he graduated he was offered the post of teacher of mathematics and astronomy in Graz, a small provincial capital in Austria. This was in part a recognition of his prodigious mathematical skill, already evident, and in part a result of his somewhat iconoclastic religious attitudes, which made the priesthood an unsuitable profession for him (he remained a Lutheran throughout his life, but he never accepted the complete orthodoxy). Later, when he had become an astronomer famed throughout Europe, his unconventional religious attitudes encouraged the belief among the Jesuits that he might convert. This would have simplified his life in that period of intense religious turmoil, for he lived in a Protestant minority often persecuted by the Catholic majority. However, his intellectual honesty prevented him from embracing a doctrine foreign to his beliefs. By his own admission, he was a poor teacher. Writing in the third person, in his horoscope, he confesses that his "enthusiasm and eagerness is harmful and an obstacle to him" because he thinks of "new words and new subjects, new ways of expressing or proving his point, or even of altering the plan of his lecture or holding back what he intended to say." Kepler was now in his middle twenties, and, as has been so often the case among the great scientists, it was out of this period of youthful enthusiasm and flexibility of mind, which probably seemed to be confusion to his students, that the fundamental guidelines to his most important scientific ideas emerged.

Kepler's early scientific career is especially interesting because the ideas that seemed to *him* to be the most significant, and which he tried to exploit for the rest of his life, appear to a modern reader to be almost completely mad. It was the fact that he could never get them to work that drove him to make the discoveries that appear to *us* to be so significant. From the beginning, he was convinced that the basic astronomical verities must have a geometrical interpretation. This conviction has been shared by all the great natural philosophers, from Pythagoras to Einstein—the conviction that the cosmos was laid out according to a mathematical design and that this design is "simple" and accessible to human intelligence. For Kepler, mathematics meant the pure geometry of the Greeks. God was for him a master Greek geometer, and the "book of the world" must therefore be contained among the theorems of Euclid. One of them stated that there are only five "perfect solids." A perfect solid (the most familiar example is the cube) is a solid all of whose faces are "perfect" plane figures. (In the cube, these figures are squares.) The other perfect solids are the tetrahedron, the octahedron, the dodecahedron, and the icosahedron. There were known to be six planets—Mercury, Venus, Earth, Mars, Jupiter, and Saturn, in order of increasing distance from the sun, around which, Kepler believed, the planets moved in circular orbits. Carrying on with his geometry, he considered a universe in which a cube, a tetrahedron, a dodecahedron, an icosahedron, and an octahedron would be arranged concentrically, one inside another; the orbit of Mercury would be fitted within the first of these perfect solids, the orbit of Venus outside it, and outside each of the other solids the orbit of another planet. This, he thought, might make it possible to calculate the interplanetary distances and also explain why there were no more than six planets. "Why waste words?" he wrote in his *Harmonice Mundi* (*Harmony of the World*). "Geometry existed before the Creation, is co-eternal with the mind of God, *is God himself.*"

With the superior vision of hindsight, it is all too easy for us to pass judgment on the weakness of Kepler's youthful notion. (Apart from anything else, we know now that there are *nine* planets.) In fact, if Kepler's mysticism had not been coupled with a fanatic obsession to make his theory fit the observed facts quantitatively, he might well have gone down in scientific history as just another visionary crank, along with the more unenlightened alchemists who abounded at that time. (It is interesting to note that Newton also devoted his "spare" time to alchemy.) This combination of mysticism and devotion to the celestial "facts" as he knew them was Kepler's great strength. Einstein characterized the interrelation between mystic intuition and the need to deal with the hard facts in the formula that "Science without religion is lame. Religion without science is blind." Kepler was quite aware of the dangers of pure speculation conducted without taking into consideration the facts learned through observation: "I play, indeed, with symbols, and I have started a book called 'The Geometric Cabbala,' which is concerned with reaching the forms that I *am* playing. For nothing is proved in symbols alone; nothing hidden is arrived at in natural philosophy by geometrical symbols . . . unless it is shown by conclusive reasons to be not only symbolic but to be an account of the connections between things and their causes." Kepler published his fantastic scheme, at the age of twenty-five, in his first book, *Mysterium Cosmographicum* (*The Cosmographic Mystery*), in 1596, when the anti-Copernican attitude among the religious Establishments, both Catholic and Protestant, was hardening. Armitage notes that

> It had been Kepler's original intention to begin his book with a frank avowal of his belief in the Copernican theory, and to maintain that it was not inconsistent with Scripture to assign the central place to the sun. But he yielded to the counsel of the Tübingen authorities, with whose approval the work appeared, and agreed to treat the theory merely as a mathematical hypothesis leading to interesting results. By adopting this course he insured that the book should be judged on its scientific merits only and that believers should not be scandalized. "The whole of astronomy," he wrote, "is not worth so much that one of Christ's little ones should be injured."

The then current data on the planetary orbits, primarily supplied by Ptolemy and Copernicus, were not very accurate, and Kepler was able to work out a rough agreement between these orbits and his own chimera—an agreement that satisfied him at least for the moment. In the *Mysterium* he included a section on the role of the sun in determining the planetary motions. Up to this point, it had apparently not occurred to anyone, even the Copernicans, to ask *why* the planets revolve around the sun; i.e., what is the role of the sun? Kepler was impressed by the fact that the planets move more quickly in their orbits when they are close to the sun and that the nearer planets move more quickly in their orbits than do the exterior planets. His intuition told him that something emanating from the sun must account for this. At first, he thought that the planets might possess a "motive soul," which the sun agitated when the planets came closer. By the second edition of the *Mysterium,* twenty-five years later, the phrase *anima motrix* ("motive soul") had been replaced by *vis* ("force"). Kepler's idea of what this force was kept changing. For a time, he thought that it might be the pressure of the light of the sun. He was able to demonstrate that the intensity of light diminishes with the square of the distance of the observer from the source. (Coincidentally, as Newton later postulated in his *Principia,* so does gravitation, which, of course, is the real force, exerted by the sun, that causes the planetary motions.) However, this explanation did not satisfy Kepler, for if it were accurate the planetary motions, notably the motion of the earth, would be interrupted during a solar eclipse. In 1600, the English scientist William Gilbert published his classic text on magnetism. Kepler read it intently, and he came to believe that the sun exerted a magnetic force that moved the planets. Moreover, at about this time, it was discovered that the sun rotates on its axis. Kepler tried to relate the sun's rotation to that of the planets. Late in his life, in the notes to *Somnium,* his science-fiction work (the notes grew to be much longer than the book itself), he describes a concept of universal gravitation that resembled in its basic outlines the ideas that Newton made precise a half century later. But Kepler was prevented from fully understanding the gravitational mechanism because of his faulty notion of inertia, which he had acquired from Aristotelian physics.

In 1597, Kepler, now in Graz, married Barbara von Mühleck, already twice widowed at the age of twenty-three, whom he later described as "simple of mind and fat of body." Although it lasted fourteen years, until her death, it was hardly an idyllic marriage. To make matters worse, the persecution of the Lutherans began in earnest. Koestler notes, "The young Archduke Ferdinand of Hapsburg (later Emperor Ferdinand II) was determined to cleanse the Austrian provinces of the Lutheran heresy. In the summer of 1598, Kepler's school was closed down, and in September all Lutheran preachers and schoolmasters were ordered to leave the province within eight days or forfeit their lives." Kepler went briefly into exile, but was allowed to return for a time, probably because of the influence of important members of the Jesuit order who felt that he might still be converted. (To one of them he wrote, "I am a Christian, the Lutheran creed was taught me by my parents, I took it unto myself with repeated searchings of its foundations, with daily questionings, and I hold fast to it. Hypocrisy I have never learnt. I am in earnest about Faith and I do not play with it.") But in 1600 he and his family were forced to leave Graz, and they went eventually to Prague.

Nevertheless, Kepler's life and career were now marked by an extraordinarily fortunate coincidence, involving one of the most bizarre figures in the history of science, the Danish astronomer Tycho Brahe, who was born in 1546 into the Danish nobility. At the University of Copenhagen, he fought a duel with a fellow student, apparently over which of them was a better mathematician, and lost part of his nose. It was replaced by a false nose, of gold and silver alloy, and he carried a snuffbox filled with an ointment that he incessantly rubbed on it. His lifelong preoccupation with astronomy began at the university when he witnessed a partial eclipse of the sun. What impressed him was not so much the eclipse but that it had been predicted to occur exactly when it did. (Foretelling eclipses is a skill that goes back at least as far as the Babylonians.) After three years in Copenhagen, Tycho (as he is usually known) was sent by an uncle to the University of Leipzig in the company of a young tutor, Anders Soerenson Vedel, who, Koestler reports, later became the first great Danish historian. Vedel was instructed to dissuade his charge from further astronomical study, a career unfitting for a nobleman. Tycho, however, carried along some small astronomical instruments, which he hid under a blanket and used after his tutor had gone to sleep. At the end of a year, the tutor gave in and Tycho began openly studying the heavens. Koestler reports:

> Tycho continued his studies at the Universities of Wittenberg, Rostock, Basle, and Augsburg until his twenty-sixth year, all the time

collecting, and later designing, bigger and better instruments for observing the planets. Among these was a huge quadrant of brass and oak, thirty-eight feet in diameter and turned by four handles—the first of a series of fabulous instruments which were to become the wonder of the world. [The telescope was not invented until after Tycho's death, in 1601.] Tycho never made any epoch-making discovery except one, which made him the father of modern observational astronomy, but that one discovery has become such a truism to the modern mind that it is difficult to see its importance. The discovery was that astronomy needed *precise* and *continuous* observational data.

On the eleventh of November, 1572, Tycho, who had returned to Denmark and was skywatching full time, on the estate of another uncle, and was now completely familiar with the celestial patterns, observed something in the constellation of Cassiopeia that caused him to doubt his own vision. Indeed, he called his servants and several peasants to confirm the phenomenon. He had witnessed the first explosion of a star, or nova, to be recorded since 125 B.C., when the Greek astronomer Hipparchus witnessed a similar event. Until modern times, it was believed that the phenomenon Tycho had witnessed was the creation of a new star (*nova* is the Latin for "new"), but it is now believed that a nova is a stupendous thermonuclear explosion of an existing star—the most spectacular way in which it can end its career. Tycho's nova, which at its brightest outshone the planet Venus and could be seen during the day, was visible for eighteen months, then became too dim to be perceived by the naked eye. It was witnessed throughout Europe and created tremendous excitement, for both scientific and extra-scientific reasons. Its appearance was considered to have great astrological portent, and according to Koestler, the German painter George Busch, who claimed that the nova was a comet condensed from the rising vapor of human sins and set afire by the wrath of God, felt that the fallout from the burned debris would cause all sorts of evil, such as "bad weather, pestilence, and Frenchmen." The question of whether the nova was actually a comet, a planet, or a star was settled by Tycho, who, making careful observations, discerned that its position appeared to remain constant, so it could be assumed that it was a fixed star, not a planet or a comet. At any rate, the natural philosophers were faced with the fact that a new object had been created in the stellar domain, which the orthodox cosmologists, following Plato and Aristotle and their subsequent interpreters, had felt was immutable. Remarkably, two more nova-like objects were sighted in the next thirty-odd years—in 1600 and 1604. The last followed closely a rare triple coincidence of Mars, Jupiter, and Saturn.

In 1576, King Frederick II, in recognition of Tycho's growing reputation, offered him, as an observatory, the entire island of Hveen, which lay north of Copenhagen, in the strait between Denmark and Sweden. He renamed it the Island of Venus, and on it he built an observatory castle, which he called Uraniburg. Koestler's description of the scholarly routine at Uraniburg makes a modern scientist feel that science has lost a certain *on ne sait quoi:*

> Life at Uraniburg was not exactly what one would expect to be the routine of a scholar's family, but rather that of a Renaissance court. There was a steady succession of banquets for distinguished visitors, presided over by the indefatigable, hard-drinking, gargantuan host, holding forth on the variations in the eccentricity of Mars, rubbing ointment on his silver nose, and throwing casual tidbits to his fool Jepp, who sat at the master's feet under the table, chattering incessantly amidst the general noise. This Jepp was a dwarf, reputed to have second sight, of which he seemed to give spectacular proof on several occasions.

For twenty years Tycho remained on his island, compiling the most accurate data on the planetary orbits that had ever been assembled and ruling despotically over his subjects. But King Christian IV, Ferdinand's successor, finally decreased Tycho's perquisites, and in 1597 Tycho departed, with his retinue. In 1599, after traveling through Europe, they arrived in Prague, where Tycho accepted the post of Imperial Mathematicus under the auspices of Rudolph II and where he was eventually to meet Kepler.

Tycho and Kepler were well acquainted with each other's work. Kepler wanted to use Tycho's planetary data in order to fill out the details of his chimerical perfect-solid universe, and Tycho wanted Kepler's mathematical genius—already made evident in the *Mysterium,* a copy of which had been sent to Tycho—in order to perfect his own chimerical scheme. But it was not the custom of the times to make scientific data available to rivals, and Kepler wrote Mästlin to say of Tycho:

> Any single instrument of his costs more than my and my whole family's fortune put together. . . . My opinion of Tycho is this: he is superlatively rich, but he knows not how to make proper use of it as is the case with most rich people. Therefore, one must try to wrest his riches from him.

Tycho had devised a universe in which the planets went around the sun and the sun went around the stationary earth. It was a compromise between Ptolemy and Copernicus, designed to placate the schoolmen. Tycho, however, was having difficulty in making it jibe with what he had learned from his orbital sightings, and in 1599 he invited Kepler to come to Prague and join him in astronomical study. Kepler was banished from Graz some months later, but he had already accepted the invitation, and thus fate arranged for the two great founders of modern astronomy to meet just when they needed each other the most.

In February of 1600, Kepler took up residence in the castle of Benatek, which was twenty-two miles from Prague and which Tycho was making over in the image of Uraniburg. He was already having financial difficulties with the emperor, who was beginning to exhibit the symptoms of mental instability that finally led to his downfall. Tycho had lost some of his assistants, and Kepler was assigned the task of determining the orbit of Mars. Tycho's senior assistant, Longomontanus, had been struggling with the Martian orbit for several years, but its complexities had baffled him, and he was switched to studying the moon. Kepler bet that he could determine the Martian orbit in a week. It took almost eight years—years that were documented in his magnum opus, *A New Astronomy, or A Physics of the Skies.* He simply had not realized what it meant to cope with data as accurate as Tycho's, and he wrote that the problems they raised "took such a hold of me that I nearly went out of my mind." Kepler occupied a social position halfway between Jepp and the Master, and Tycho would part with only bits and pieces of his data as if, Koestler writes, "he were handing bones to Jepp under the table," and Kepler wrote a friend that "Tycho gave me no opportunity to share in his experiences. He would only, in the course of a meal and in between conversing about other matters, mention, as if in passing, today the figure for the apogee of one planet, tomorrow the nodes of another." Tycho may have suspected that Kepler wanted to use the data for his own purposes and that once he had them he might abandon his work on Tycho's system. As it happened, Tycho died a year and eight months after Kepler's arrival, and Kepler, whom Rudolph appointed Tycho's successor, a position he occupied for the next twelve years, spirited the data away from Tycho's heirs.

In 1609, *A New Astronomy* appeared in print. Of it, Armitage comments:

> This is not a treatise, a systematic presentation of results, but a testament, the record of an almost spiritual pilgrimage, conducting the reader along all the windings of the road (and up all its blind alleys) and recording the play of the great astronomer's passing moods. With its intricate calculations and speculative flights of fancy, the "New Astronomy" is the most difficult to read of all the half-dozen decisive cosmological books of the world, and the more so as its author was wrestling for much of the time with mathematical problems requiring for their rigorous and elegant solution concepts and notations not at that period available.

What is most significant about the book has come to be summarized as the first and second of Kepler's laws of planetary motion. The first law states that all of the planets move in elliptical, not circular, orbits around the sun and that the sun is not in the center of the planetary ellipses but at their foci. (There is nothing at the center of the ellipses.) Kepler, studying the orbit of Mars, first abandoned the idea that the planet moved at uniform speed in a circle and then the idea that it moved in a circle at all. It was fortunate that he had been given the Martian orbit to resolve, since, apart from Mercury, whose orbit was not well known because of the planet's proximity to the sun, which obscures it, Mars (of all the planets discovered at that time) was the one whose orbit exhibited the greatest, and therefore most apparent, "eccentricity." Finally, Kepler realized that the Martian orbit is an ellipse. This discovery, one of the most pivotal in the history of science, did not arouse much enthusiasm. Koestler quotes a letter to Kepler by David Fabricius, a clergyman and amateur astronomer:

> With your ellipse you abolish the circularity and uniformity of the motions, which appears to me the more absurd the more profoundly I think about it. . . . If you could only preserve the perfect circular orbit, and justify your elliptic orbit by another little epicycle, it would be much better.

Even Galileo could not accept the concept of elliptical orbits. It was Newton who at last realized that an elliptical orbit is natural, given the form of the law of universal gravitation. But for Kepler there was no framework into which he could fit his discovery, and all his life it was to him more a "cabin" than a "cathedral."

Kepler's second law deals with how planetary motion is affected when planets approach the sun. For this purpose, he resurrected an enormously complicated and lengthy method of computing areas devised by Archimedes. But he was thus able to verify his second law, which describes quantitatively how the rates of motion of a planet are related to each other at different points on the orbit—the so-called law of equal areas in equal times. (That is, if one were to chart the progress of a planet, as it revolves around the sun, by drawing an imaginary line from the center of the sun to the center of the planet, one would discover that the area traversed by that line in, say, the course of any given hour is equal to the area traversed by the line in any other given hour, no matter how elliptical the orbit of the planet, though the two areas, because the orbit is an ellipse, might be of quite different shapes.) It predicts in particular that the planets will accelerate when they approach the sun. Again, it took Newton to show how this law can be derived from the properties of the gravitational force. For good measure, in the introduction to *A New Astronomy,* Kepler gave the first correct explanation of oceanic tides on the earth—that they are caused by an attractive gravitational force exerted by the sun and the moon. *A New Astronomy* is, altogether, an almost incredible display of scientific invention.

Tycho's heirs, enraged by Kepler's appropriation of the planetary data, managed to delay the publication of *A New Astronomy* for a time. The Junker Tengnagel, who had married Tycho's daughter Elisabeth and who had been made an appellate counselor at court, managed by a legal maneuver to attach a preface of his own manufacture to the book. "Greetings to the reader!" he announces. "I had intended to address thee, reader, with a longer preface. Yet the mass of political affairs which keep me more than usually busy these days, and the hasty departure of our Kepler, who intends to leave for Frankfurt within the hour, only left me a moment's time to write. But I thought nevertheless that I ought to address a few words to thee, lest ye should become confused by the liberties which Kepler takes in deviating from Brahe," and on and on. The book was printed in 1609 in a beautiful folio volume, of which only a few copies survive. Ironically, soon after *A New Astronomy* appeared, Kepler's life in Prague became all but unbearable. As Armitage tells it:

> The year 1611 proved one of the most tragic in the annals of Kepler's chequered life story, and it marked the break-up of his career at Prague. Yet he began the year cheerfully enough with the dispatch of a seasonable offering to his patron and fellow-countryman Wackher von Wackenfels. It took the form of a disquisition, half playful, half scientific, on the geometry of snow crystals. . . . As the tragic year 1611

drew on, Kepler's home circle was shattered by the deaths, first of his promising six-year-old son, Frederick, and then, in July, of his wife, Barbara. Meanwhile, civil war had reached Prague. The Emperor Rudolph had completely lost his grasp of the political situation. His cousin Leopold, Bishop of Passau, bringing up an army supposedly in support of Rudolph, had occupied part of Prague against local resistance, and when these forces had been bought off, Archduke Matthias [Rudolph's brother] was left in control. Rudolph was compelled to abdicate and Matthias became King of Bohemia and, in due course, Emperor. Efforts were made from several quarters to involve Kepler in the struggle in his capacity of an astrologer able to discern which of the contending parties might expect to be favoured by the celestial influences. While scrupulous to avoid giving comfort to his master's enemies, Kepler was equally careful not to encourage the Emperor in any imprudent venture.

Kepler hoped to find an academic position in Tübingen, but the authorities, doctrinaire Lutherans, were suspicious of his Calvinistic leanings. After Rudolph's death, Matthias kept Kepler on in his position as Imperial Mathematicus, and he received permission to go to Linz, where he was to live for the next fourteen years. He had already thought of going to Linz, before his wife's death, because of her increasing melancholia, which he believed might be stemmed were she to return to her native Austria. Armitage remarks, "After the social and intellectual life of Prague, Kepler must have found Linz something of a backwater. . . . Once again, as in Graz, he found himself a mere assistant master, subordinated to the petty tyrant of a district school." And Kepler's unorthodox Lutheran views led to his being expelled from the local parish church. (He did find a congregation, outside Linz, that would allow him to take communion.) In 1613, Kepler, by now forty-one, married Susanna Reuttinger, who was twenty-four and who had been chosen by him after a long consideration of a dozen candidates, whom, characteristically, he describes in meticulous detail in an eight-page folio letter. Susanna gave Kepler, for the first time, a happy homelife. They had six children, of whom three survived infancy, and she took care of the two surviving children of Kepler's first marriage. Armitage writes:

> Kepler was a devoted father, zealous in the education and religious instruction of his children. He prepared German versions of Latin texts for the use of his son Louis, and he composed a catechism on the Sacrament for the children to commit to memory.

Indeed, Kepler might have settled into a serene middle age if the political and social atmosphere of the times had allowed it. However, in 1615 he began one of the worst ordeals of his life—the witch trial of his mother, Katherina. Witch-hunting had become a mania in Germany at the beginning of the seventeenth century. "In Weil-der-Stadt, Kepler's idyllic birthplace, with a population of two hundred families, thirty-eight witches were burnt between 1615 and 1629. In the neighboring Leonberg, where Kepler's mother now lived, a place equally small, six witches were burnt in the winter of 1615 alone," according to Koestler. However bleak our contemporary age may sometimes appear, the atmosphere of superstition, ignorance, and cruelty that the descriptions of Kepler's ordeal evoke make one appreciate our era of at least relative enlightenment. Kepler's mother, nearly seventy, was a meddlesome crone. Her friend Ursula Reinbold, the wife of the town glazier, had become pregnant because of an extramarital escapade and had undergone an abortion to escape scandal. She confided in Kepler's mother, whose interminable gossiping made the affair public. Ursula countered by saying that Katherina had gotten her pregnant by casting an evil spell. Ursula's husband and his brother offered all sorts of corroborative evidence. A lame schoolmaster recalled that his disability had followed his taking a drink from Katherina's tin cup, and another neighbor testified that his wife had withered and died after taking a drink from the same cup. The wife of the butcher swore that one of her husband's thighs became inflamed when Katherina passed by, and the tailor said that Katherina was responsible for the deaths of his two children because she had said some incantations over their cradles in the guise of blessing them. Soon the whole town was able to find Katherina's hand at the root of its misfortunes, and, Lear points out in his preface to the *Somnium:*

> Katherina unfortunately accommodated her detractors by continuing her restless rounds of unwanted advice. Worse still, she kept urging her homemade remedies upon everyone for every sort of indisposition, always from the now notorious tin cup.

The situation so deteriorated that Kepler's relatives, who had exacerbated it by suing Ursula for slander and thus forcing her to collect even more evidence, wrote to him for help—the first hint he had of his mother's difficulty. He wrote to the officials of Württemberg demanding documentation of the charges, and, Lear observes, "thereby discovering that he himself also had been accused of 'forbidden arts.' " Kepler replied by reminding the town council of Leonberg that he was his Roman Imperial Majesty's Court Mathematicus and demanding that all the documentation be sent to him. But now Kepler's mother attempted bribery of an official, thus seeming to confirm her guilt. She fled to Linz, where she remained for nine months, then returned to the home of her daughter in another town in Germany. Kepler followed, passing the journey by reading *Dialogue on Ancient and Modern Music,* by Galileo's father, and received permission to take his mother back to Linz, but she refused to go. Then followed, Koestler reports, "a strange lull of two years—the opening years of the Thirty Years War—during which Kepler wrote more petitions and the court collected more evidence, which now filled several volumes." His mother was arrested in 1620, and he set out once again for Leonberg. He applied all of his energies in his mother's defense, the effect of which was summarized, Koestler notes, "by a slip in the court scribe's record: 'The accused appeared in court, accompanied, alas, by her son, Johannes Kepler, mathematician.' " Nevertheless, after fourteen months of confinement, she was released into his custody. She died six months later. Koestler adds, "It was against this background that Kepler wrote 'Harmony of the World,' in which the third planetary law was given to his engaging contemporaries." We shall come back shortly to this law.

The leitmotiv of Kepler's scientific work is sounded in his own words: "Thus God himself was too kind to remain idle and began to play the game of signatures signing his likeness unto the world: therefore I chance to think that all nature and the graceful sky are symbolised in the art of Geometria." No scientist has ever better expressed the underlying faith of *all* scientists that the world is "comprehensible" (Einstein's word). *Harmony of the World,* completed in 1618, is the last of Kepler's great original scientific works, but until his death, in 1630, in Regensburg, he continued his enormous production of books, tracts, and letters, including a brief tract on the method of logarithms, which he needed to simplify his numerical computations—*Epitome Astronomiae Copernicanae,* which, despite the modesty of its title, is really a summary of Keplerian astronomy; practically the only point of similarity between his and the Copernican system is that both men held that the earth was in motion around the sun—and finally, his *Rudolphine Tables.* This work, upon which Kepler had labored on and off for twenty-six years—it was finally published in 1627—became the standard astronomers' handbook for the next century. It contained, among other items, a list of the positions of the known fixed stars, tables of planetary orbits, logarithm tables, and the longitudes of the most important towns of the world. (The original edition contains a frontispiece depicting the Temple of Astronomy supported on ten pillars, five of which represent one or another of the great creators of astronomy whose trophies they bear aloft. Kepler modestly occupies one of the carved faces of the base. He is seated at his study table scribbling on the cloth, as if to suggest that he is too poor to buy writing paper, though the Imperial Eagle is dropping him coins.)

But *Harmony of the World* contained the last of Kepler's monumental contributions to astronomy—the third law.

The third law states that the ratio of the square of a planet's period (that is, the time it takes to make a circuit of the sun) to the cube of its distance from the sun is the same for all the planets in the solar system. Each planet moves in an ellipse, and for most planets the ellipse is almost a circle, which is why the concept that the planets moved in a perfect circle in the course of their orbits survived as long as it did. Kepler chose Saturn to illustrate his law. The distance from earth to the sun is about 93 million miles, and Saturn is about nine times that far away from the sun. Since the earth requires a year to go around the sun, Saturn should, by Kepler's law, require twenty-seven years. As a matter of fact, it takes thirty years, but that is because only *approximate* distances are used in this particular calculation, and if the exact distances are used Kepler's law is verified. If all this seems complicated to us today, one can imagine how it must have appeared to Kepler's contemporaries. This law was buried almost casually, it would seem, in *Harmony of the World,* alongside a vast assortment of observations about music and astrology that, from our point of view, are at best historical curiosities. Kepler found his third law because he was looking for it. His concept of God the geometer convinced him that there must be a relationship between the distances between the planets and the sun and the time it takes the planets to go around the sun, and he simply experimented with all the relationships he could think of until he found the one that worked out. He did not, and could not, know why it worked. It was a half century before Newton demonstrated that Kepler's third law was not only consistent with the mathematical form for the gravitational force the sun exerts but actually implies that the law of gravity has the mathematical form that it does. Without in any way diminishing Newton's tremendous achievements, it is not too much to say that Kepler's three laws were the fundamental building blocks on which the

Newtonian synthesis was erected. Kepler's law presented Newton with a summary of the empirical facts, which he then translated into a well-defined problem in mathematical physics; i.e., to find the law of force that would compel the planets to move in ellipses. Ellipses are "conic sections," and a conic section is a curve created when a cone is bisected by a plane. The curve that results, depending on the angle of the plane to the base of the cone, is a circle, an ellipse, a parabola, or a hyperbola. Planets move in ellipses, a ball thrown from the surface of the earth moves in a parabola, and comets, at least those that escape the solar system, move in hyperbolas. It was, again, Newton who demonstrated that the gravitational attraction of the sun requires that the comets move in conic sections, like the planets, and moreover, that planetary motion obeys the second and third Keplerian laws as well. The law of Newtonian gravitation states that the sun's gravitational force acting on any planet is proportional to its mass and the mass of the planet and inversely proportional to the square of the distance from the planet to the sun. Kepler came close to discovering this for himself, but he didn't. Newton was the first physicist to grasp the full implications of the inertia of motion, which had to a considerable extent been clarified by the work of Galileo. He also had the mathematical power to connect the law of force to the planetary orbital shapes—a problem in the integral calculus.

The career of Galileo touched to an extent on Kepler's life and work, and both Koestler and Armitage deal with Galileo, who is still a controversial figure in the history of science; there is a trend among modern historians to downgrade his character and achievements. Koestler belongs to this school and is quite frank about it. Galileo was an extremely difficult and egocentric individual. Philosophically, he was more of a positivist than a mystic, and it is likely that Kepler's mysticism and geometric visions irritated him and made him underrate Kepler's real scientific achievements. The two men never met. They did correspond, but most of the letter writing was done by Kepler. It began when Kepler, then twenty-five, sent Galileo a copy of his *Mysterium,* the book in which he announced his fantasy about the perfect solids. For Galileo, the most important aspect of the book appears to have been Kepler's public avowal of its Copernican base. Galileo, who was then thirty-three, replied that he had "adopted the teaching of Copernicus many years ago" but that he had not dared to bring his Copernican ideas into "public light" because he did not want to be ranked with Copernicus, who "is yet to an infinite multitude of others (for such is the number of fools) an object of ridicule and derision." Galileo did not acknowledge his Copernican beliefs in print until he was nearly fifty. The reason, most probably, had to do with physics as much as it did with the religious temper of the times. It simply was not possible to *prove* that the earth moved. Choosing a coordinate system in which the earth moves and the sun is at rest recommended itself solely because of its simplicity of description. However, one must realize that accepting this "simplicity" meant abandoning an entire cosmological outlook, and one which was deeply connected with prevailing religious beliefs. One can well understand why this seemed unreasonable to Galileo's contemporaries. What one condemns them for is not that they refused to accept the motion of the earth. (A modern scientist confronted with the evidence then available might not have accepted the idea, either.) One condemns them because, after Galileo's trial, in 1633, *one could lose one's life* for proclaiming that the earth moves! In all of history, it has never been possible to reverse the progression of scientific ideas. An intellectual or political tyranny may succeed in suppressing an idea for a generation, or even several generations, but it does so at the peril of losing the respect and confidence of the generations that follow.

Kepler answered Galileo's letter at once, "first because it meant the beginning of a friendship with an Italian; secondly, because of our agreement on the Copernican cosmography." He urged Galileo to declare publicly his Copernicanism, added that if he was unwilling, "Let me know, at least privately if you do not want to do it in public, what you have discovered in support of Copernicus," and proposed that he use whatever instruments were at his disposal to observe the motion of the stars. There was no response for twelve years. Meanwhile, Kepler became interested in optics, and in 1604 he published his first work on the subject, *Supplement to Witelo, Giving the Optical Part of Astronomy* (the Polish optician Witelo had, in the thirteenth century, written a basic optical treatise, *Perspective,* which Kepler felt he was bringing up to date). In his *Supplement,* Kepler gave for the first time a correct description of the function of the retina in vision and made an attempt to find a valid optical law of "refraction." Refraction is the "bending" of light when it passes from one medium to another—air to water, for example. (A stick put in a pool of clear water appears to bend where it enters the water.) Unless astronomers know to what extent the earth's atmosphere bends starlight, they cannot properly determine the positions of the stars. The law by which the amount of bending can be ascertained was devised by Willebrord Snell, the Dutch astronomer and mathematician, during Kepler's lifetime, but not published until seven years after he had died. Kepler, however, discovered a reasonable approximation to Snell's law—one that worked for air and water and enabled him to come close to making the necessary corrections in the estimates of the stellar positions. So, with his interest in optics, he was fully prepared to understand the telescope, when it was invented, and the use that it was put to by Galileo for astronomical discovery. It is probable that the telescope was invented around 1608 by the Dutch spectacle-maker Johann Lippershey. Galileo, hearing about it, manufactured one for himself. In 1610, he published *Sidereus Nuncius (The Star Messenger),* in which he announced his observation of the imperfections in the moon's surface (this was important in breaking the hold of Greek science, since at this time it was widely believed, as the Greeks had believed, that the moon was a "perfect" body, like the planets), his discovery of numerous stars, and, most important, his discovery of four of Jupiter's moons, which he named the Medicean stars, in honor of Cosimo de Medici. Galileo's book—only twenty-four leaves in octavo—had an immediate impact on his contemporaries because of its readability, as compared to *A New Astronomy* and Copernicus's *Book of the Revolutions,* and because it opened up the awesome prospect that the contents of the universe might be inexhaustible—an idea that the modern mind may find exhilarating but that an age still very much under the influence of the "closed" medieval cosmology found extremely disturbing. There was an almost violent negative reaction to *The Star Messenger,* and to this Galileo characteristically responded in kind. One reason for the reaction was that with the first telescopes it was hard for an untrained observer to see what Galileo insisted was there. Kepler, now the Imperial Mathematicus, and a man to be reckoned with in the intellectual community, saw a copy of *The Star Messenger,* and was sent a request from Galileo for an opinion. Though he had not heard from Galileo in twelve years, and though he did not yet have a telescope of his own with which to confirm Galileo's discoveries, Kepler at once, in the hope that Galileo would reciprocate by commenting on his own *New Astronomy,* wrote a pamphlet, *Conversation with the Star Messenger,* expressing his confidence in Galileo's observations. Galileo widely circulated its contents but offered no immediate acknowledgment to Kepler. But after many efforts at refutation of Galileo's work had been made by astronomers who had telescopes but were unable to find the Medicean stars, he wrote Kepler, thanking him for his support and promising new telescopes "to my friends" (presumably Kepler was one) so that they could confirm the Galilean statements. Kepler never heard again from Galileo, and he got word of Galileo's later telescopic discoveries, announced in all but undecipherable Latin anagrams, only through intermediaries. He also never got one of Galileo's telescopes. He did succeed in borrowing one and confirming for himself the existence of the Medicean stars. One can only guess what the subsequent history of science might have been if Galileo and Kepler had collaborated. As it was, Kepler was spurred on because he was able to use a Galilean telescope, and in his classic book, published in 1611, *Dioptrics,* he initiated the science of geometrical optics by tracing, theoretically, the path of a light ray through a telescope and explaining how it works. He also invented what came to be known as the Keplerian telescope, which had convex lenses instead of a concave and a convex lens, and which gave the viewer an upright rather than an inverted image.

Kepler seems to have begun writing *Somnium (The Dream),* in 1609, almost as soon as he had discovered that the planetary orbits were ellipses. He returned to it in 1621, and in the decade that preceded his death he added 223 footnotes, which dwarf the text. It was published by his impoverished family four years after his death, in the hope that the income from it would help to sustain them. Apart from being a great work of science fiction, it is a review of Kepler's life seen in a wonderfully distorted dream. Kepler's mother appears as the enchantress Fiolxhilde. (The first draft of *The Dream* was written before her trial, and there is evidence that it was used by her enemies to document their charge that the Keplers had supernatural powers.) Kepler appears as her son Duracotus; his errant father appears as her fisherman husband, who "died at the very old age of one hundred and fifty years (when I was three) after about seventy years of marriage." Duracotus is given to a sea captain as compensation for a pouch of "herbs and patches of embroidered cloth," which, out of curiosity, Duracotus had cut open, causing its contents to be scattered to the winds. The captain carries a letter from the Bishop of Iceland, where Fiolxhilde lives, to be delivered by Duracotus to Tycho, in Denmark. Duracotus learns astronomy by watching Tycho and his students passing "whole nights with wonderful instruments fixed on the moon and stars. This reminded me of my mother because she, too, used to commune constantly with the moon. Thus by chance I, who came from very impoverished circumstances in a half-barbaric land, achieved an understanding of the most divine science, which has prepared the way for me to greater things." Duracotus, returning to Iceland, finds his mother "delighted in the knowledge I had acquired about the sky. She compared my reports of it with discoveries she herself had made about it. She said she was ready to die since her knowledge, her only possession, would be left to her son and heir."

It is at this point that the real purpose of *The Dream* begins to become apparent—Fiolxhilde symbolizes Ignorance and Duracotus is Science. "So long as the mother, Ignorance, lives, it is not safe for Science, the offspring, to divulge the hidden causes of things; rather, age must be respected, a ripening of years must be awaited, worn out by which, as if by old age, Ignorance will finally die." *The Dream* was the means by which Kepler could, in the guise of fiction, divulge "the hidden causes" without exposing himself to slander and persecution. *The Dream* was actually a brilliant popular scientific exposition of Copernican cosmology. Kepler's device is to imagine what the moving earth looks like viewed by the inhabitants of the moon, who, naturally, claim that their planet is stationary. Fiolxhilde summons forth the Daemon from Levania (Kepler informs the reader that *"lebana,"* or *"levana,"* is the Hebrew word for "moon"). Speaking in Icelandic, the Daemon describes the voyage to the moon:

> No inactive persons are accepted into our company; no fat ones; no pleasure-loving ones; we choose only those who have spent their lives on horseback, or have shipped often to the Indies and are accustomed to subsisting on hardtack, garlic, dried fish, and such unpalatable fare. Especially suited are dried-up old crones, who since childhood have ridden over great stretches of the earth at night in tattered cloaks on goats or pitchforks.

The trip is to take place during a lunar eclipse, and the Daemon goes on to describe the launch:

> We congregate in force and seize a man of this sort; all together lifting him from beneath, we carry him aloft. The first getting into

> motion is very hard on him, for he is twisted and turned just as if, shot from a cannon, he were sailing across mountains and seas. Therefore, he must be put to sleep beforehand, with narcotics and opiates, and he must be arranged, limb by limb, so that the shock will be distributed over the individual members, lest the upper part of his body be carried away from the fundament or his head be torn from his shoulders. Then comes a new difficulty: terrific cold and difficulty in breathing. The former we counter with our innate power, the latter by means of moistened sponges applied to the nostrils. . . . We entrust the bodies to the empty air and withdraw our hands. The bodies roll themselves together into balls, as spiders do, and we carry them almost by means of our will alone. Finally, the corporeal mass heads of its own accord for the appointed place.

This last sentence suggests the modern concept of inertial motion, but Kepler makes it clear that what he had in mind was the fact that at some point in the voyage the moon's gravitational attraction would take over and pull the astronaut to its surface, which shows that Kepler had understood that gravity is a universal phenomenon. What he had not discovered was that an object in motion will keep moving even if *no* force is acting on it. According to Newton, the earth, as well as the sun and the other planets, exerts a gravitational attraction on the moon. Why, then, doesn't the moon fall toward the earth? After all, if one simply drops an object in midair, it falls toward the earth. However, if one gives the object a momentum perpendicular to the earth's surface, it acquires an inertial component in this direction, acting in addition to the force of gravity, and the object moves in a parabola, not directly toward the earth. What Newton realized was that if the perpendicular momentum is great enough, the object will go into orbit, like the moon. Throughout his life, Kepler continued to try to find what it is that sweeps the moon along and what turns the planets around the sun, and all the while the answer lay in the understanding of inertia.

Kepler's astronauts reach the moon, a land with "very high mountains and very deep and broad valleys. . . . Moreover, the whole of it . . . is porous and pierced through, as it were, with hollows and continuous caves, which are the chief protection of the inhabitants against heat and cold. Whatever is born from the soil or walks on the soil is of prodigious size. Growth is very quick; everything is short-lived, although it grows to such enormous bodily bulk." Some inhabitants "have no settled dwellings, no fixed habitation; they wander in hordes over the whole globe in the space of one of their days, some on foot, whereby they far outstrip our camels, some by means of wings, some in boats pursue the fleeing waters"—an echo of the precariousness of Kepler's own existence. "A race of serpents predominates in general. It is wonderful how they expose themselves to the sun at midday as if for pleasure, but only just inside the mouths of caves, in order that there might be a safe and convenient retreat."

Kepler lived and worked in a time of terrible political upheaval, religious bigotry, and superstition. His own "safe retreat" was in the serene contemplation of the Harmonies of the World. Koestler discovered a passage in one of Kepler's late letters that could serve as his epitaph: "Sagan in Silesia, in my own printing press, November 6, 1629. 'When the storm rages and the state is threatened by shipwreck, we can do nothing more noble than to lower the anchor of our peaceful studies into the ground of eternity.' "

2

RABI: THE MODERN AGE

ON the afternoon of March 29, 1974, there was a brief ceremony in Pupin Laboratory—the locus of the Columbia University Physics Department—in honor of the department's most distinguished professor emeritus, Isidor Isaac Rabi. A bust of Rabi, in bronze, was presented to the department, and there was a great deal of good-natured discussion about how much the bust actually resembled Rabi. (Rabi is short, perhaps five feet six, white-haired, and given to wearing dark suits.) About one matter there was no discussion, and that was how much the Columbia Physics Department, with its galaxy of past and present—and, no doubt, future—Nobel Prize winners, has reflected the personality of Rabi for nearly a half century. I. I. Rabi created the modern Columbia University Physics Department not once but twice: the first time during the 1930s, when he and a small number of his colleagues and friends, including J. Robert Oppenheimer, brought the new physics—quantum physics—back from Europe, where they had been students; and the second time after the Second World War, when the department had been brought to a standstill by the wartime projects of much of its faculty, including the development of the atomic bomb and radar. Rabi himself was then one of the prime movers in the development of radar, at the Radiation Laboratory of the Massachusetts Institute of Technology (MIT), in Cambridge. He also functioned as a trouble-shooter for Oppenheimer at Los Alamos, New Mexico, and participated in all the early decisions that physicists had to make in order to advise politicians and the military in the totally new field of atomic energy. A few years after the war, he was one of the physicists, along with Enrico Fermi, who initially opposed the construction of "the super"—an antiquated and unworkable version of the hydrogen bomb. They were overruled by President Truman, and it was Oppenheimer's association with this negative advice as much as anything else that brought about his collision with the military and led, in the atmosphere of the McCarthy era, to his security-clearance hearing in 1954. In April of that year, Rabi, who has never had the slightest fear of speaking his mind to anyone, testified at Oppenheimer's security hearing before the Atomic Energy Commission (AEC) in Washington, "There is a real, positive record. . . . We have an A-bomb and a whole series of [them]. . . . What more do you want—mermaids? This is just a tremendous achievement. If the end of that road is this kind of hearing, [I think it is] a pretty bad show."

It was Rabi's conviction, shared by other scientific leaders of the time, that an American president must have reliable scientific and technological advice—that the president must have his own science adviser within the White House. This feeling became urgent when Sputnik was launched in October of 1957 and it suddenly appeared that the Russians were about to outstrip the West in military technology. President Eisenhower had great confidence in Rabi's integrity and judgment—confidence that stemmed from an incident in 1947, before Eisenhower became president of Columbia University. During a visit to Columbia, Eisenhower made a statement that Rabi thought was totally wrong, and Rabi said so in no uncertain terms. Eisenhower never forgot this, and when he became president of the United States he sought Rabi's advice, knowing that Rabi would always say exactly what he felt, whether or not he believed that it would meet with presidential approval. When Rabi suggested that the president have a science adviser, Eisenhower agreed, and James Killian of MIT was appointed to the post. This office functioned with varying degrees of success under Presidents Eisenhower, Kennedy, and Johnson, but in 1973 President Nixon abandoned it. (It has now been restored.) Beyond all this, in the last half century Rabi has been one of the most important physicists in terms of scientific discoveries. He refuses to classify himself as either a theoretical or an experimental physicist, but his Nobel Prize, awarded in 1944, was given for a series of experiments dealing with the magnetic structure of nuclei—experiments of such elegance and coherence that the papers emerging from them, in the 1930s, have been used as the basis for courses in nuclear physics.

Rabi was born on July 29, 1898, in Rymanów, in Galicia, which belonged to the Austro-Hungarian Empire. Soon after, the family emigrated to thc United States. "My parents came to this country not as a refuge from persecution, as did the Jewish refugees from Russia," he told me during a series of conversations I had with him about his life and career. "They lived quite freely in the Jewish community of Rymanów. They had no resentment against Emperor Franz Josef. He was apparently well liked. The real problem was how to make a living, and my father, who had no skills, finally had to emigrate in order to make a living. When he left, he was quite young—he must have been twenty-one or twenty-two—and he went by himself to America. He got himself established, and a few months later was able to get some money to send for my mother and me."

I asked Rabi what language he had learned first. "Yiddish," he answered. "Yiddish, which I really spoke very well—like a native. My parents spoke Polish occasionally. It was a secret language, to keep things away from the children, but otherwise we spoke Yiddish."

Like so many other Jewish immigrants, the Rabis first lived on the Lower East Side of New York. "I had to learn English in the streets," Rabi told me. "I remember one time—one doesn't really remember how one learns a language, but I remember one time asking another boy, 'What is the English word for *zucker?'* And he said 'sugar.' And after a while—I suppose by the time I was four or five—my English was very good; that is, for Lower East Side English." Rabi's parents were deeply religious. "They had a very strong Orthodox, Chassidic, fundamentalist tradition, which they carried on in New York," he recalled. "I was raised in that, and heard all the stories that go with it—all the stories about evil spirits and devils, and, of course, on the other side, God and the Founding Fathers, so to speak. For those people, God was present all the time. In conversation, not a paragraph—hardly a sentence—would go by without a reference to God. The Orthodox Hebrew religion requires a lot of doing; there is a lot of ritual, so you're pretty busy at it. It isn't just going to the synagogue on Saturdays and holidays. Indeed, if you have the time, if you don't have to make a living, you go there every day—three times a day, as a matter of fact. You're kept pretty busy with prayers and with the learning that accompanies them. It is not a superficial kind of culture."

To make a living, Rabi's father began doing such jobs as delivering ice until he "graduated into work in the sweatshop, making women's blouses." Eventually, he borrowed some money and opened a grocery store on the Lower East Side. A few years after the Rabis had settled here, Rabi's sister, who is five years younger than he, was born. They were the only children of the family. "Both my mother and my father came from large families—twelve or thirteen—and they saw the evils of a large family," Rabi said. "They wanted to educate their children."

Rabi told me that as far back as he could remember he never

doubted that he would go to college. "When I look back at it now, it seems absurd, because we had hardly enough money for food," he said. "All the tales that you read about those times—I am talking now of 1900 to 1907, when we lived on the Lower East Side—all the tales you hear about people living in slums, that's the way we lived. Still, I never had any idea I was living in a slum. Nor did I have any envy of richer people—they were just richer, we were just poor. That's all there was to it. So there was no class consciousness of the sort that you see in people now." The neighborhood was divided up according to the residents' place of origin in Europe—Russian Jews with Russian Jews, Galician Jews, like the Rabis, with other Galician Jews, and so on. Every attempt was made to preserve these boundaries—including attempts to prevent intermarriage among the different Jewish communities. "It was very densely populated—lots of kids around," Rabi said. "There were street gangs, and the Jewish gangs were very tough. There were holdups. They didn't have policy games, quite, but something like them. There were lots of prostitutes and saloons—every intersection had three or four saloons. And, of course, there were a lot of synagogues. There were no Christians living there then. Some of the synagogues were, so to speak, fossil churches—they had been bought and 'cleansed' by an elaborate ritual to convert them into synagogues."

I asked Rabi at what age he had begun to go to school, and he said, "It was the tradition in my family that a boy started going to school at the age of three, which I did. My reading was in Yiddish books and Bible stories. Because I was very small, I needed protection. I had it, because I was so good at telling Bible stories. I'd keep them fascinated—the big boys. I was kind of a mascot because of my Bible stories."

In those early years, Rabi was rather sickly. (He now appears to enjoy robust health.) "I had every kind of disease that was going," he told me. "You know, children's diseases. In that congested region, I caught them all. I really kept my family poor with doctors' bills. Well, as we look back on it now, it is a wonder that people survived there." The food was pretty poor, as Rabi remembers it—"very little protein." There were no government social services. "If my father was out of work, he was out of work, and there was nothing coming in," he recalled. "The cupboard was bare." There were voluntary neighborhood associations, some religious and some consisting of people who had come from the same European town or village. "There was one to which my father belonged," Rabi said. "It was called the Rymanów Young Men. These were societies in which, among other things, one could borrow small sums. There was a tremendous amount of self-help within the community."

Rabi noted that the great virtue of being brought up in such crowded conditions was that the children were exposed from their earliest years to all manner of serious problems and issues. "You heard a lot of conversations on all sorts of topics," Rabi told me. "Most of them you didn't understand, but you understood something. Listening to them did develop my vocabulary, and it gave me a familiarity with the basic issues in life, which a middle-class child doesn't get at all anymore. Now the children are put to bed early, and the real conversation goes on when they're not there, whereas in my family we were right in the midst of it. There was hardly an important issue of the times that I didn't hear about and hear discussed—very passionately, too. It was really so superior to what middle-class children have now, especially if they live in a place like Scarsdale. They see nothing but greenery and have to be taken places to find somebody to play with."

When Rabi was about nine, his family moved to Brownsville, in Brooklyn—which was then almost suburbia. There were chickens in the streets, and Rabi can remember going to buy milk from a local farmer. He continued his education at P.S. 156, in Brownsville. (His first public school had been P.S. 22, on the Lower East Side.) "The New York school system was at that time wonderful in many respects," Rabi remarked. "For example, it took these immigrant boys and turned them into Republicans. We became very patriotic." There was rigid discipline in the public schools. Rabi remembered, some seventy years later, a time when the students in his class were supposed to be lined up for some reason. "I was a bit out of line, and the teacher slapped me in the face and knocked me over into a trough," he told me. "I remember being hit over the knuckles with a ruler a few times, too. But in general things were all right. I did well in school, but I was no prodigy. Neither did I do any work. I went to class and listened, and I read outside—but not the classwork." It was in public school that Rabi learned of English-language books. At home, the only books he had ever seen were prayer books in Hebrew and Bible stories in Yiddish. Then, in 1908, he discovered the local branch of the Brooklyn Public Library. "I saw some kid in class who had a book that was clearly not a schoolbook. I asked him where he got it, and he told me about the library. I went and got a library card and took out two books." The librarian stopped him and made him read from them out loud. Rabi was so small and looked so young that the librarian simply could not believe he knew how to read.

I asked Rabi if he had had any particular experience in those early days that turned his mind toward science. "Yes," he said. "A very profound one. One time, I was walking along and looked down the street—looked right down the street, which faced east. The moon was just rising. And it scared hell out of me! Absolutely scared the hell out of me. Another profound experience that I had revolved around the first verses of Genesis. They were very moving to me as a kid. The whole idea of the creation—the mystery and the philosophy of it. It sank in on me, and it's something I still feel. But, as a matter of fact, I got into science in a funny way. I read all the fairy stories and other stories in the children's section of the library. I started with Alcott and worked down through Trowbridge—all those children's books, all those writers. Then I came to the end of *those* shelves, and there was a science shelf. So I started with astronomy. That was what determined my later life more than anything else—reading a little book on astronomy. That's where I first heard of the Copernican system and the explanation of the changes of the seasons, the phases of the moon, and the idea that the stars were suns, very distant suns. Ours was such a fundamentalist family that my parents hadn't heard of the Copernican system, and for me it was a tremendous revelation. I was so impressed—the beauty of it all, and the simplicity."

I asked Rabi if his reading had led to a religious crisis with his family. "Yes, I think there was a kind of religious crisis," he answered. "It came gradually. You see, when you are a fundamentalist everything is covered by religion. As you get a wider perspective, bit by bit, there are contradictions. Besides, the Orthodox life demands so much. Since I was always inclined to be lazy, my religion gradually began to slip away. My father used to say, 'It's not that you're irreligious—you're lazy.' But the biggest contradiction was with the Copernican theory. There is certainly a contradiction between that and what I was told as a child—that the earth is flat and that there is a big fish, the Leviathan, which surrounds it, which has its own tail in its mouth, and which will be eaten on the Day of Judgment by the good Orthodox Jews while the others will get nothing. It's a bit of a jump from there to the Copernican system. Besides, when you are Orthodox you say prayers for the new moon. When you have the astronomical explanation, the rising of the moon becomes a sort of non-event. My family was very conservative. Even Maimonides was a bit suspect. In fact, *more* than a bit suspect—a great man, but still. . . ."

Rabi paused, and then went on, "When I think about it, for my family it was rather tragic, because there was nothing they could do. I was the firstborn, the only son. And I could outtalk them. They hadn't read about Copernicus. So we reached a sort of modus vivendi where at home I conformed to everything. I didn't try to persuade them of anything else. They would have really suffered from that. And they didn't ask very much what I did outside. So I was a good son in that respect. I tried to help with the holidays—to make them joyful occasions. I have a great respect and a great feeling for those things. It's part of a culture, a way of life, an outlook. Sometimes I feel I shouldn't have dropped it so completely—I'm talking about the way of life. There's no question that basically, somewhere way down, I'm an Orthodox Jew. In fact, to this very day, if you ask for my religion, I say 'Orthodox Hebrew'—in the sense that the church I'm not attending is that one. If I were to go to a church, that's the one I would go to. That's the one I failed. It doesn't mean I'm something else."

I told Rabi, whom I have known for nearly twenty years, that I had always sensed in him the same kind of religious feeling that Einstein seemed to have had. Rabi remarked that with Einstein there was something else as well, since when Einstein said things like "God would not play dice"—meaning that the laws of physics would not be allowed by God to be fundamentally statistical—he was, Rabi felt, identifying himself, to a certain extent, with the Creator. "Einstein *was* a world-maker," Rabi said. "In any event, I think that 'God' is a very good heuristic principle—a standard by which you can judge things. The idea of God also helps you to have a greater feeling for the mystery of modern physics. I don't mean the way some of the young people I see, who are very good, take physics. They take it as a system you can do things with, can calculate something with, and they miss—at least, so it seems to me—the mystery of it: how very different it is from what you can see, and how profound nature is. We could not have arrived at modern physics just by extrapolation from what we already knew. It had to result from discovery—from real discoveries. I have never thought of physics as a profession, and in that sense I think I'm a gentleman as far as science is concerned. I never thought of doing it as a living. I didn't care what I did for a living. But science was something I would follow and admire. And enjoy. I think that most of the people I knew who studied physics in my time, who were more or less my age, didn't have this attitude, quite. They were doing it as something that they liked to do. Some said, 'It's fun.' I always hated the idea that it was 'fun.' I know other ways to have fun. Physics has a much deeper emotional quality for me than that."

Getting back to Rabi's public-school days, I asked him if he realized then that science was a profession. "I knew that there was an 'astronomer,' but a scientist as such, no," he answered. "Engineering, yes. I thought of becoming an astronomer. I told my father I might become an astronomer, and he said, 'How do you make a living out of this?' I had no idea. I thought you did it by teaching, or something like that. I didn't understand the difference between science and engineering, or know that physics existed and could really be a profession." When Rabi was ten, he began reading about electricity in the local library, and soon he was immersed in telegraphy—both standard and wireless. By the time he was eleven or so, he had learned the Morse code, had both a sending and a receiving station, and had taken some of the telegraphy examinations that are required by federal regulations if one has a transmitting station. "I organized my own group, and we set up a telegraph system that went over a few blocks," Rabi told me. "I got the equipment, and finally discovered a boy whose father was a junkman. We got wire from him and strung it across the street, so high that it was hardly visible. We had a neighborhood telegraph system." Barely into his teens, Rabi published his first scientific paper, which dealt with making a condenser—a device for storing electrical energy. The paper was published in one of the late Hugo Gernsback's numerous magazines—*Modern Electrics*. Gernsback, who founded *Amazing Stories,* the first science-fiction magazine, sent Rabi a reprint of the article a decade ago. He had paid two dollars for the original manuscript.

In addition to carrying on his technological activities, Rabi continued to read intently. "I used to be able to read with extraordinary rapidity," he told me. "I could read a big book a day. I can't now. I can't read a tenth of a book a day. But at that time I'd come home from the library at the beginning of the week with about five books, and at the end of the week they were back." A book that made a deep impression on him then was Jack London's *The Iron Heel,* published in 1907. "I don't know whether you know it," Rabi said. "It describes, among other things, a workers' revolt in Chicago, set in what was then the future—about 1930. In it, he gave, in just a few chapters—beautifully written, as I remember it—a précis of Marxian theory. He gave the theory of class struggle, the labor theory of value, the theory of surplus value, which makes revolution inevitable, and the materialistic interpretation of history. It made an enormous impression on me. I went on to read other authors, like John Spargo, and tried to convert people—fellow students, even some of the teachers—to socialism. For a while, I took it very seriously. I went to the meetings of the Socialist party in the neighborhood, and all that. After a while, I began to have the feeling that the Socialists were either kidding themselves or trying to kid me. Part of the Socialist thing was 'equality': anybody can be president, anybody can do this or that; it just depends on the circumstances of your life. But after I went to high school and looked at my classmates, I said, 'Those people can't run a government or a world,' and dropped the whole thing. But it was an interesting background for me—especially the materialistic interpretation of history, since it was an actual theory of history. When I studied history, all of this fell into place. It wasn't just an arbitrary set of dates. For me, history became a living thing, and later on in high school, when we started taking the Regents Examination for the State of New York, I made the highest grade in history in the whole state. It was just a natural for me. I still have a great interest in history and in its interrelatedness."

After Rabi finished grade school, his parents had wanted him to go on with Hebrew studies—preferably at a yeshiva. Rabi refused. "I was through with that, partly thanks to Copernicus," he told me. Instead, he elected to go to Manual Training High School, in Brooklyn, which he describes as "a second- or third-rate high school." He explained, "I went there purposely. Normally, I should have gone to Boys High—that was *the* school. All the very smart Jewish kids went to it. But I had been raised in an environment where we didn't see anything but Jews. My elementary school was just full of Jews, with maybe here and there a few gentiles who had, as it were, crossed district lines. I wanted to get away from that. I had very definite notions of being an American in a broader sense. I had read a good deal of history, and I wanted to be part of the greater thing, so I went to Manual Training, where there were almost no Jews. I had four years of shopwork—woodworking, cabinetmaking, printing, and various kinds of mechanical work. I did that sort of thing at home with wireless telegraphy and model airplanes. But the great advantage of going to a second-rate school was that I didn't have to do more than ten minutes of homework a day."

I asked Rabi if his interest in science had continued while he was at Manual Training. "I was interested, but not in the school science," he answered. "My physics teacher was one of the worst. He was the only teacher there with a Ph.D., too. In fact, all my life I never got good instruction in physics. The chemistry courses were the strongest, and I took a lot of chemistry. Our math teacher was marvelous, and there was, in addition, history, of course—American and European history. American history is so badly done. They give it in an elementary way three times. You just keep on getting it."

Rabi's math teacher was also the high-school football coach, and this fact determined Rabi's choice of college. He and a friend were asked to help sell tickets to the high-school football games. "For which service we got complimentary tickets," Rabi told me. "This was important to me, because I had no money. I began going to the games and became interested in football. Now, this was a period when Cornell was preeminent in football. It was in the days of Charlie Barrett, who was their All-American quarterback. So I became interested in Cornell. It had a greater emphasis on science and engineering than other places had. There seemed to be a liberal attitude about it, a sense of freedom and novelty. Furthermore, it was out in this romantic country—romantic to me, at least, as a reader of Fenimore Cooper. That whole storied Finger Lakes Region. Here I was, a boy from the city. It seemed like a nice place to go."

Considering Rabi's general financial situation at the time, I wondered how he had managed to get enough money for college. "I had a New York State Regents scholarship—a hundred dollars a year, which was a lot more money then than it is now," he said. "I also had a tuition scholarship to Cornell, which was won in an open competition, and the rest I got from home—very little money. As a matter of fact, I lost several teeth because of malnutrition the first year. I didn't know about nutrition, and it seemed silly to me to spend money on vegetables when I was hungry and needed more solid food. And I didn't want to work. A lot of students supported themselves—they'd work in cafeterias and other places. But I didn't want to work; I preferred to starve, and I did."

Rabi began at Cornell in 1916 as a student of electrical engineering. "It was pretty dull stuff, pretty bad," he says now. His roommates were taking chemistry, so he began taking chemistry—skipping the elementary course, the material of which he already knew from his reading. He told me, "What got me excited about chemistry was a course that most people didn't like—qualitative analysis. One was given an unknown and was supposed to find out what was in it. There was a definite way of doing it, and I thought it was wonderful—like research. But basic theoretical chemistry was not known then at Cornell. In physics, Einstein's theory of relativity was not well-known anywhere in America, let alone at Cornell. My freshman year was the first time that conduction electricity had been taught there through the theory of the electron." In 1906, the English physicist J. J. Thomson had been awarded the Nobel Prize in physics for discovering the electron—which he had done in 1897, some twenty years prior to the period that Rabi was discussing. "In lectures that I gave a little later, I would often say that I came, essentially, from an underdeveloped country as far as science was concerned. Just as in most underdeveloped countries, there were a few great men, but an undergraduate in America would never see them and could not be influenced by them."

Rabi finished his undergraduate program at Cornell in 1919, after only three years. At that time, the students had been mobilized for the war. "I wore a uniform—the Student Army Training Corps—and I was living in a barracks and marched and did KP. I would have been in the war, without question, if it hadn't stopped." His degree was Bachelor of Chemistry. I had always supposed, until I asked him, that Rabi had gone right on to graduate school. "No," he said. "I thought, I'll have to make some money and then go into science. I was obsessed at that time by the difficulty of finding a job. I wasn't offered any jobs, while the non-Jewish boys were all snapped up by various chemical companies."

I asked Rabi if there was a good deal of anti-Semitism in the big chemical companies at that time. "I don't know," he answered. "But Jews didn't get jobs in them, or in the universities. It was really terrible. Well, I did find a job, but it was a menial sort of thing. I was a chemist for Lederle Laboratories, in New York, analyzing mother's milk and furniture polish and things of that sort. And then, with a friend of mine, I worked in a sort of private banking thing—discounting accounts receivable. Then he and I ran a local newspaper in Brownsville—the Brownsville *Bulletin.* I was by that time almost twenty-four."

In 1922, three years after Rabi had received his bachelor's degree, he decided to return to school. "It was now or never," he told me. "So I went back to Cornell with the idea of doing graduate work in chemistry. By then, I was interested in physical chemistry. They had no physical chemistry, so I thought I would take some physics. I already knew the chemistry, and I thought I could combine the two. So I started taking physics, and I realized that the part of chemistry I liked was called physics, so I took advanced courses in physics. I had a very hard time following them, but I did it. I worked very hard at it." In the summer of 1923, Rabi decided to transfer to Columbia, partly because he had failed to get a fellowship he had applied for at Cornell and partly because he had just met his wife-to-be, Helen Newmark, who was from New York. "It was in the summer session at Cornell. She had just come up to Ithaca from New York, more or less for the hell of it, with another girl, and I happened to meet her." He and Helen Newmark were married after he had had three years at Columbia—on August 17, 1926, the day after he sent off his doctoral dissertation. To support himself at Columbia, Rabi took a job at City College as a tutor. It paid $800 a year and required him to teach sixteen hours a week, during the day. "After the first year, I taught both day and night," he said. "It came to about twenty-five contact hours a week—at least four times the load that physicists get now. I don't know how I did it. I didn't seem to be working hard. Nobody saw me working hard. I was doing my dissertation, and even found time to go to the opera—you know, standing room. Maybe my students were badly taught, since I was not single-minded about teaching, but we got through."

At that time, many physics departments had no theoretical physicists as such and very little modern physics at all. Rabi remembers a visit to Cornell in 1922 by the great German physicist Arnold Sommerfeld. "I was sitting in the students' library, and I would see our professors sneak in to take a look at Sommerfeld's book *Atombau und Spektrallinien—Atomic Structure and Spectral Lines*—just to have something to talk with him about," Rabi recalls. Columbia graduate students, in effect, went shopping for a thesis. They would go from instructor to instructor until they found someone who would take them on and give them a problem to do. First, Rabi went to Professor Bergen Davis, who was an X-ray man. But Davis had only a certain number of X-ray machines, and they were already taken up by other students. Then he went to Professor A. P. Wills, who specialized in magnetism. "He suggested measuring the magnetic susceptibility of sodium vapor," Rabi told me. "This was a terrifically difficult experiment to do. Sodium at high temperature is, of course, a gas, and it is corrosive and very hard to handle. Well, I thought about it for a while, and finally I said to myself, 'I'm not going to do this.' So I went to him and told him that I didn't think I was good enough to do it. I couldn't tell him I didn't *want* to do it, and, besides, I didn't think I *would* be able to do it, and I didn't know anyone else at that stage who could do it, either. Wills said that he was disappointed and that he would give the problem to someone else. I said to myself, *'Moichel!* Good riddance.' Wills had no other problems, so I was essentially on my own. Just at this time, the Nobel Prize-winning British physicist William Lawrence Bragg came to Columbia to give a colloquium. He had just measured the electric susceptibility of a whole group of crystals—the Tutton salts, such as nickel-ammonium sulfate-hydrate and potassium-sulfate hexahydrate—which have similar crystalline structures but very different electric and magnetic properties. Bragg's results looked interesting, so I said to myself, 'Why shouldn't *I* measure their magnetic susceptibility?' I went to Wills. He said all right, and that was it. I didn't get any help from him. Nor did I want any. The last thing in the world I wanted when I was doing my dissertation was for somebody to help me."

The standard method of measuring the magnetic susceptibility of crystals at that time had been invented by Woldemar Voigt in Germany in the late nineteenth century. It required cutting accurately three separate sections of the crystal—each section oriented in a different direction within the crystal. The sections were then suspended on a torsion balance in a magnetic field, and the degree of positive or negative response of the crystal to the magnetic force was measured. These three measurements gave enough information to reveal the magnetic structure of the crystal, which varied in different directions within the crystal. All this required a great deal of hard work and the skill of "a good watchmaker," Rabi recalled. "I started right off to grow the crystals, which was very nice," he went on. One makes a solution of the material to be crystallized and then inserts a small piece of the crystal, around which the solution accretes to produce, or grow, a large crystal. "I could just leave them and go off to the opera," he said. "This, of course, was the trivial part. But there came a time when you could no longer temporize, and had to face the very difficult job of cutting and grinding the crystals. We had no equipment for that, and the whole business of measuring the magnetic field looked pretty tedious. Being lazy, as my father said, I just couldn't face doing the experiment the standard way."

Always a tireless reader, Rabi happened at that time to be going through one of the treatises on electricity and magnetism by James Clerk Maxwell, the nineteenth-century Scottish-born physicist, whose monumental work created the modern theory of electromagnetism. Rabi's eye was caught by a passage that discussed the force of a non-uniform magnetic field on a body immersed in a medium—a fluid with a magnetic salt dissolved in it, for example. ("Non-uniform" means simply that the magnetic field has different strengths in different regions of space.) If a medium has a certain magnetic susceptibility, he read, and if one immerses in the medium an object that itself has a magnetic susceptibility, then the object's effective magnetic susceptibility is altered by its being in the medium. If the medium has a greater susceptibility, then the immersed object will move in a certain direction in a non-uniform magnetic field; if the medium has a smaller susceptibility, the object will move in the opposite direction; and if the susceptibilities are the same, there will be no motion. This was all that Rabi needed to know. He made a saturated solution that would not dissolve the crystal he was interested in, attached the crystal to a piece of glass fiber, immersed it, and turned on a magnet. He could now vary the magnetic susceptibility of the medium by adding to it known amounts of a standard magnetic salt. When a certain amount was added, the effective susceptibility of the crystal vanished—something that Rabi could observe, since at that point his crystal would not move in the magnetic field produced by his magnet. Rabi then drew out some of this fluid in a tube and weighed it in a magnetic field. He compared this weight with that of an identical tube of water weighed in the same magnetic field. This gave him the susceptibility of the crystal compared to that of water, which was known. (Since the susceptibility varies in different directions in the crystal, Rabi could find the entire directional variation by rotating the crystal in the solution and checking the response of each section to the magnet.) His method was so simple that he was able to do an entire class of crystals—the monoclinic isomorphic series of double sulfates—and three additional crystals of another kind, which was more than had ever been done before, and to do them with better accuracy. The whole thing, from the time Rabi began seriously applying Maxwell's ideas to his experiment until he finished his measurements, had taken him about six weeks.

I asked Rabi if Wills had been impressed. "He must have been," Rabi answered. "Everybody knew that such measurements were very difficult, and so I got a good deal of credit out of it as a great experimenter. Actually, it gave me no experience in experimental techniques at all, because all I needed was a magnet, a switch to turn the magnet on and off, a good chemical balance, a blowtorch to melt the glass out of which I drew the glass fiber, and some shellac to attach the fiber to the crystal. It was something that you could do without taking your jacket off. I never was a real experimental physicist. I had no experimental technique whatever. The first time I picked up a galvanometer, I broke the suspension. I had to fix it. I never was an experimental physicist—or a real theoretical physicist, either. It's always embarrassing when people ask me, 'What are you?' I say, 'I'm a physicist.' 'Are you an atomic physicist?' 'No.' 'Are you a solid-state physicist?' 'No.' 'Are you a nuclear physicist?' 'No. I'm just a physicist.' " Rabi dissolved in laughter at the idea that someone nowadays could describe himself as "just a physicist." At present, the American Physical Society, which is the professional society of American physicists, has broken the discipline of physics down into ten technical divisions, each with an innumerable number of subdivisions, both experimental and theoretical.

Rabi went on with reminiscences of his dissertation. "The old method of measuring the magnetic susceptibility of crystals required great skill, but it was crude," he said. "It was a head-on attack—there was nothing really clever about it. It didn't have what the Germans call a *Witz*. There's no good translation of a *Witz*. It's a joke or a trick. It's the use of this kind of witty trick that I have always liked about physics. I think physics is infinite. You don't have to try to exhaust it in your generation, or in your lifetime. It's a human sort of thing, and the important thing is to do it in an interesting way, with a certain style. You see, I have always taken physics personally. I like it better that way. It's my own physics, within my powers. It's between me and nature. Nature's inexhaustible—nobody's going to take it away. All my experiments, I think, had that quality of *Witz* about them. I wasn't interested in doing them if I couldn't do them in that way. I wasn't interested in a brute-force way of doing them—unless a problem was of such interest and its solution was in such demand that I just couldn't wait for a result, or unless I was up against intense competition. Then you bull your way through. Otherwise, I have always felt that physics was the only occupation for a gentleman. I am not talking about applications—rather, pure physics. You are addressing nature in its most profound form. And you do it with a certain kind of style, a kind of intellectual style."

The year 1926, when Rabi wrote his dissertation, was about the time the modern quantum theory was being developed by such men as the German physicist Werner Heisenberg and the Austrian physicist Erwin Schrödinger. Previously, there had existed what is now referred to as the old quantum theory—basically, a set of ad-hoc recipes for quantizing physical systems; that is, assigning sets of allowed values to certain parameters (such as energy) associated with these systems. The procedure had been invented by Niels Bohr in 1913, when he postulated that only certain electron orbits around the proton—"the Bohr orbits"—were allowed in atomic hydrogen. (Atomic hydrogen, the simplest element of all, consists of a heavy proton as its nucleus with an electron circulating around it. The Bohr orbits are the allowed trajectories for this electron.) This postulate enabled him to deduce the atomic-hydrogen light spectrum, and his result agreed with earlier experiments—a fact that lent some credence to the scheme. (In Bohr's picture, the electron "jumps" from orbit to orbit, emitting light quanta as it does so. Since electrons in these orbits are allowed to have only certain energies, the light emitted can have only certain allowed frequencies—"colors"—and this accounts for the nature of the atomic-hydrogen light spectrum.) The whole procedure of assigning these orbits appeared rather arbitrary, however, and, besides, there were a growing number of instances where it either failed to give any prediction about atomic phenomena or gave the wrong prediction. With the Heisenberg and Schrödinger formulations of quantum mechanics—they were shown to be equivalent—these old recipes could be given a rational basis when they were valid and could be suitably amended when they were not. In the Schrödinger picture, for example, one describes the electron in the hydrogen atom by a probability function—the Schrödinger wave function. This function is large at the positions of the old Bohr orbits. Hence it is highly probable that the electron will follow one of these orbits, which is why the Bohr picture worked—at least approximately.

Rabi was reading the quantum-mechanics papers from Europe as fast as they arrived at Columbia, and soon he and another student—Ralph Kronig, who went on to become a distinguished theoretical physicist—decided that they would do an exercise based on the new theory. "Kronig and I said, 'That's new stuff. Let's try to do something.' " They got out Max Born's *Vorlesungen Über Atommechanik*—an encyclopedic work on atom mechanics—to see what had already been done. They found that the "symmetrical top"—a solid shaped like a top—had not yet been treated in the new wave mechanics. They set up the Schrödinger equation—the equation that determines the Schrödinger wave function—which was a fairly heavy piece of applied mathematics, especially for the time. (It was also done at about the same time by two well-known German mathematical physicists, Fritz Reiche and Hans Rademacher, in collaboration.) Recently, Rabi told me some of the circumstances surrounding his calculations with Kronig. They found themselves confronted with a hostile-looking differential equation, which neither of them knew how to solve. One day, Rabi was prowling through the General Library at Columbia when he came on a book by the great nineteenth-century German mathematician Karl Gustav Jakob Jacobi (1804-51). He began leafing through it, and suddenly he said to himself, "My God! That's our equation!" It turned out that Jacobi had come across it and solved it in another context. Rabi and Kronig then published a paper in *The Physical Review,* a journal published by the American Physical Society. One feature of Rabi's and Kronig's solution is that it exhibits the quantum character of angular momentum and of energy. In quantum mechanics, such a top can have only a discrete set of values for its angular momentum and energy. These values correspond to the various quantum states of the system, and *these* correspond to different solutions of the Schrödinger equation. As the top being considered becomes more macroscopic—corresponding to the classical-physics situation—these energy and angular-momentum levels come closer together, so a classical top, while quantized in principle, has energy levels so close together that the quantum effects are essentially unobservable, and the same thing is true of the angular-momentum levels. Rabi's and Kronig's work has now become a standard textbook example in quantum theory.

In August of 1926, Rabi sought some sort of fellowship that would take him and his bride to Europe, for he wanted to work with the quantum mechanicians. There were at the time a few fellowships available to Columbia graduate students—two Cutting Fellowships and the Barnard Fellowship. "A Cutting Fellowship was a very good one, and the Barnard was very small," Rabi told me. "It provided $1,500, out of which you had to pay for your travel. I applied for a Cutting Fellowship but didn't get one. I did get the Barnard, so Helen and I traveled fourth class, when it existed. When it was abolished, we traveled third class. We never stayed in a third-class hotel, though—always something lower. Youth hostels and that sort of thing. When we finally got to Germany, we lived the way German students lived, and we got to know Germany in that way—including the language."

At present, anyone who wants to study at an established European scientific institution has to apply formally and go through an elaborate screening process. In those days, none of this seemed to be necessary. Rabi simply decided where he wanted to go and showed up there, unannounced. "As I look back on it, it seems remarkable—the way I was received," he told me. "Nothing special, but it was not considered unusual that I just appeared like that. Since I was interested in quantum mechanics, I went first to Zurich, where Schrödinger was. He was just about to leave, so I stayed only a couple of weeks. But it was there that I first met some of the people—like Julius Stratton, who became president of MIT, and Linus Pauling—who are now among my oldest friends. Pauling was on a Guggenheim, and Stratton was getting his doctorate at the Swiss Federal Polytechnic School [Einstein's alma mater] in Zurich. When Schrödinger left, I went to Munich, where Arnold Sommerfeld was teaching, and stayed until the end of the semester. There I met people like E. U. Condon, the theoretical physicist, who died not long ago, and H. P. Robertson, the mathematical physicist. Hans Bethe, who won the Nobel Prize in physics a few years ago, was a graduate student."

After the semester ended, the Rabis spent a few days in England, and then went on to Copenhagen, where, seven years earlier, Bohr had established his Institute for Theoretical Physics.

"We arrived in Copenhagen from England and checked our bags," Rabi recalled. "I bought a map, and we walked to the Institute. I rang the bell. A secretary answered. I said, 'My name is Rabi; I've come to work here.' Again, we had arrived without previous arrangement. She gave me a key to the Institute and the name of a *pension* where Helen and I could stay. I left Helen at the *pension,* went back, and started to work—nobody around. This for days. Then somebody showed up. It turned out to be Jordan." This was the German physicist Pascual Jordan, who later, in collaboration with Max Born and Werner Heisenberg, made fundamental contributions to the quantum theory. "I had trouble making him out at first, because he stammered so badly. All the others arrived within a few days, and I was working away at my thing, but not very effectively. I was trying to compute the magnetic susceptibility of a hydrogen molecule. I was hipped on magnetism. Now, for some reason or other, Bohr was very tired at that time, and so the people around him arranged, without telling me, for me to go to Pauli, in Hamburg." Wolfgang Pauli was one of the most formidable of the quantum theorists. "Well, as it turned out, getting 'thrown out' of Copenhagen was the greatest thing that ever happened to me."

In Hamburg, there were two principal physicists: Pauli and Otto Stern. Both greatly influenced Rabi's subsequent career. Rabi told me that before he went to Europe he had never come across a first-rate mind. "And there in Hamburg I was close to the greatest minds of the time. Yet I discovered that I knew a lot more physics than the Germans of my age and training—fresh Ph.D.s. I knew a lot of physics, but I would describe my knowledge by saying I had the libretto without the score. In other words, I was not yet immersed in the living tradition of physics. A lot of what you really accomplish in physics depends on your taste in what to work on and what to be interested in. You can work very hard on an unimportant problem and show tremendous skill and ingenuity. Before I went to Europe, I had never met real producers of physics. And such physics, and at such a time! It was the most formative experience I had."

Pauli, who died in 1958, was born two years after Rabi, and was only twenty-seven when Rabi met him. He had been a celebrated child prodigy—and one of the few who have gone on to accomplish something in science in adult life—and just out of his teens he had published what is still one of the best critical monographs on the special and general theories of relativity. He was a devastating but fair critic of scientific ideas, and he had a total contempt for mediocrity. He once characterized a young physics professor whose work he thought was particularly bad with the remark *"So jung und schon so wenig geleistet"* ("So young and already he has contributed so little"). If an idea was really terrible, Pauli would say that it wasn't even wrong. Rabi told me that the young Pauli "was what Pauli always was—very tough," and went on, "But if you got the first blow in, it was all right. Interestingly, I managed to get the first blow in. We were talking about something, and I wanted to disagree, but my German wasn't very good, so I said, *'Das ist Unsinn'*—'That is nonsense.' One didn't say that to Pauli—he kept walking around afterward saying, *'Unsinn? Unsinn?'* Later, it turned out that the disagreement was purely linguistic—the similarity of 'p' and 'pi.' In German, 'pi' is pronounced like 'p.' It was a misunderstanding. But he never attacked me the way he did some of the others. We got along very well, especially later on, when he came to the United States." If there ever was someone to learn taste from in the choice of scientific problems, it was Pauli; and another advantage for Rabi in working with Pauli was that many of the famous quantum mechanicians, including Bohr, came to Hamburg frequently for visits during that period, to use Pauli as a sounding board for their ideas.

The really decisive influence on Rabi's scientific career, however, was Otto Stern. Stern, Rabi recalls, was not as impressive on first contact as Pauli, but one came to realize that he was "very profound." Stern was born in Sorau, Upper Silesia, in 1888. He moved to Breslau as a child, and in 1906 he entered the University of Breslau to study physical chemistry. In 1912, he received his doctorate. Around that time, he came to realize—just as Rabi did later—that the part of chemistry he liked was physics. In Stern's case, it was Einstein's work—which was by then quite widely known among European physicists—that really attracted him to physics. After receiving his degree, he moved to Prague, where Einstein held a professorship for a brief period, and in 1913, when Einstein became a professor at the Swiss Federal Polytechnic School, Stern followed him to Zurich. There Stern became a *Privatdozent,* or unpaid lecturer, at the same school. He began by doing various kinds of theoretical studies in the old quantum theory, but in time his interest shifted to experimental physics, and he worked in this field after moving to the University of Frankfurt, in 1914. By 1919, Stern had become engaged in the work that led to his being awarded the Nobel Prize for physics in 1943, a year before Rabi received it.

Stern's work evolved from a development of what is called the method of molecular beams. A molecular beam is simply a group of molecules moving in one direction in a vacuum, and so avoiding all but a very few collisions, either among the molecules in the beam or with gas molecules. (In any "vacuum" there are always some remaining gas molecules, but so few as to make collisions with the beam molecules unlikely.) The beam can be produced by having a furnace heat up an element—silver, for instance—so that individual molecules boil off into a vacuum. A system of slits is placed in the vacuum, and the molecules pass through these slits. The slits serve to collimate the molecules into a narrow beam. Such a gaseous beam of molecules is an ideal tool for studying the properties of individual molecules.

The first applications of Stern's method were relatively modest. He confirmed some predictions made by the classical statistical mechanics of how heated molecular gases would behave. This was excellent work but his most significant molecular-beam experiments were carried out in the period from 1920 to 1922—

during which he wrote a number of papers on his experiments, in collaboration with Walter Gerlach, a Frankfurt colleague. The experiments began routinely. Stern wanted to measure the magnetic strength of a silver atom by putting the atom in a magnetic field and studying its motion. In Stern's day, there was thought to be one basic source of magnetism—moving electrical charges. Einstein had shown in his theory of relativity that electric and magnetic fields were really aspects of the same phenomenon—electromagnetism—and that which aspect would be manifested depended on the state of motion of the charges relative to the observer. (This had been implicit in Maxwell's theory of the electron field, but its full significance was clarified by Einstein.) Now, an atom consists of a positively charged nucleus around which circulate the relatively light electrons. The moving electrons form an electric current, and this current generates a magnetic field. The magnetic field resembles, in its effects, a tiny bar magnet in which the north and south poles are perpendicular to the plane in which the electrons are circulating. Stern's idea was that he would make a beam of silver atoms and run the beam through a bar magnet—all this in a vacuum. (In this case, the molecular beam became an atomic beam. The silver atom is remarkable in that all its electrons except the one farthest away from the nucleus are locked in a tightly bound configuration. The circulation of the single external electron produces the current that generates, in first approximation, the silver atom's magnetic field.) Like any bar magnet, the one that Stern constructed, and through which he sent his silver atoms, had a north pole and a south pole. The magnet was bent to resemble the letter "C," and he shaped the metal poles so that one was essentially flat while the other was honed to a knife edge. The gap between the two pole pieces was of the order of a millimeter; the edge of the "knife," along which the silver atoms ran, was a few centimeters. This shaping produced a strong non-uniform magnetic field, which exerted a strong force on the silver atoms—a force of interaction between Stern's magnet and the intrinsic magnet of the silver atoms. The force acted perpendicular to their direction of motion. Hence, the atomic beam would be displaced from its original direction by the magnetic force. The displacement could be measured by first turning off the magnet—an electromagnet that could be switched on and off—and allowing the silver atoms to deposit themselves on a polished surface (glass, in the original experiment), and then photographing the surface.

Stern and Gerlach expected the silver atoms, when they came out of the furnace and formed the atomic beam in the vacuum, to have their own magnetic poles aligned more or less at random. When the atoms came in contact with the electromagnet, Stern reasoned, they would move in various directions. The undeviated beam deposits itself on the glass in roughly a straight line, and they expected the deviated beam to deposit itself as a sort of smear, reflecting different motions. Instead, they found something that appeared to be—and was, from the point of view of classical physics—completely incomprehensible: The beam had split in two. In their original paper is a photograph of the undeviated beam deposited on the glass—a fine line—and, beside it, a photograph of the deviated beam, split in two. (The split is a fraction of a millimeter.) There are no atoms that have passed through the system undeviated. It is as if the electromagnet had forced each silver atom to decide which of two beams it belonged in—as if every atom had to make the choice. This was the phenomenon of "space quantization." But why, and why two beams?

Stern's and Gerlach's experiments raised a question they couldn't answer, and neither could anyone else. Classical physics had simply broken down. Now we know the answer, and the knowledge has led to a profound change in outlook—to the whole new world of quantum physics. The answer is related to the quantization of angular momentum. An electron circling a nucleus has an angular momentum. The angular-momentum vector points in the same direction—in the case of an atom like silver, along the same line—as the north and south poles of the silver magnet. The strength of the atom's magnetic field is measured by what physicists call its magnetic moment (for convenience, these magnetic moments are expressed in units of "the magneton," even though no nucleus or particle has a magnetic moment of precisely one magneton), and the magnetic-moment vector and the angular-momentum vector are proportional to each other. Classical physics predicts that if such an angular momentum is measured it can point in any direction. But in quantum physics—after the measurement—it can point in only a finite number of directions. In quantum physics, angular momentum can have only the values 0, 1/2, 1, 3/2, 2, 5/2, 3, 7/2, and so on. (The numbers given here are in units of $h/2\pi$, where h is a universal constant—Planck's constant, named after Max Planck, who discovered the quantum at the turn of the century. The experimental value of $h/2\pi$ is $1.0545887 \times 10^{-27}$ erg seconds—an incredibly small unit, which is why quantum effects become crucial only on the atomic scale. Hence the "1" above stands for $1 \times h/2\pi$ and so on. One writes the expressions without the explicit appearance of $h/2\pi$ to make the equation look simpler.) The number of directions in which an angular momentum can point after measurement is $2J + 1$, J being the angular momentum in units of $h/2\pi$. This means, to take one example, that if the angular momentum is 1/2 it can point in only two directions—$2 \times 1/2 + 1 = 2$: "up" and "down," say, with respect to some direction of orientation. For a macroscopic system, like a child's top, J is a huge number, which means that, in a practical sense, the number of directions in which the angular momentum can point after measurement is essentially unlimited. When J is a small number, as in an atom, the space quantization of angular momentum is essential and must be properly taken into account to get agreement with experiment. In fact, since the magnetic-moment vectors are proportional to each other, the magnetic moment must also point, after measurement, in the same direction as the angular momentum. Thus, what Stern and Gerlach had discovered was that atomic silver had an angular momentum of 1/2, leading to two allowed directions, since $2 \times 1/2 + 1 = 2$. They did not realize this at the time. It took several years of work—inspired largely by their experiments—before the correct quantum-mechanical interpretation was worked out, and before physicists accepted the idea that a magnet like Stern's acted as a filter to project the possible states of angular momentum. However, the angular momenta of the silver atoms were oriented when, in Stern's experiments, they came out of the furnace into the vacuum, their angular momenta when they left the magnet had two, and only two, orientations—as was shown by the two separated lines that Stern and Gerlach observed.

To physicists of this period, the result of the Stern-Gerlach experiments was stunning. Rabi, who had given a colloquium at Columbia on the original experiment while he was still a graduate student, recalls the impact that it had on him. "I found the quantum theory very hard to believe—the old quantum theory—until I heard about the Stern-Gerlach experiment," he told me. "I thought the old quantum theory was stupid. I thought one might be able to invent another model of the atom which had the same properties. But you can't get around the Stern-Gerlach experiment. You are really confronted with something quite new. It goes on in space, and no clever classical mechanism would do—would explain it."

Not long after doing the experiments with the silver atoms in a vacuum, Stern moved to the University of Hamburg as a professor and the director of a new molecular-beam laboratory, which had been created specially for him by the university, and it was then that his path and Rabi's crossed. Rabi was still working with Pauli and doing other theoretical physics, but he soon began to help out with some of the theoretical aspects of the molecular-beam experiments, and it took him only a short time to arrive at an idea of how to improve the Stern-Gerlach setup. Because Stern and Gerlach were producing the spatial inhomogeneity of their magnetic field with a knife-edge magnet, the field was necessarily confined to a tiny volume of space. Moreover, the field was non-uniform, and therefore difficult to map out, and the deflections it produced were exceedingly small. Rabi proposed replacing the knife-edge magnet with a uniform magnetic field confined to a small space, and allowing the silver atoms to impinge on this field from the outside at an angle. That way, the setup would look something like a system in which a beam of light comes in contact with a square of glass, which refracts it—bends the light—at the edge. The atoms would come down the vacuum until they hit the edge of the homogeneous magnetic field. Once inside the field, they would shift in energy, and this energy shift would produce a Stern-Gerlach separation of the beam which would be larger than in the original setup. At least, Rabi thought it would be. He simply proposed the new setup as a casual theoretical suggestion, and then, much to his surprise, was asked to do the experiment. "I was invited to do it," he told me. "I had no job, and I had a wife to support. They told me it was a great honor, and I was in no position to refuse an honor. So I did the experiment—my first with molecular beams." The results were just what he had predicted.

Afterward, the Rabis moved on to Leipzig, where Werner Heisenberg was working, and there Rabi resumed his theoretical studies. It was there that he first met J. Robert Oppenheimer. Oppenheimer was born in New York City in 1904, into a family of considerable wealth. He attended the Ethical Culture School in New York, where, it turned out, he and the future Helen Rabi had been in many of the same classes, and Mrs. Rabi recalls that in the seventh grade he was already universally recognized as an intellectual phenomenon. Rabi once wrote of Oppenheimer, however, "From conversations with him I have the impression that his own regard for the school was not affectionate. Too great a dose of ethical culture can often sour the budding intellectual who would prefer a more profound approach to human relations and man's place in the universe." In any event, after graduating from Ethical Culture, Oppenheimer entered Harvard, in 1922. There he was an even greater phenomenon—graduating summa cum laude in three years with a bachelor's degree in physics. Oppenheimer went to Europe for his graduate work—first to England, where he studied for a time at Cambridge, and then to Germany, where, under the direction of Max Born, in Göttingen, he wrote a significant thesis on the quantum mechanics of molecules. It was after this that Oppenheimer came to Leipzig to be with Heisenberg.

I got to know Oppenheimer reasonably well between 1957 and 1959, when I was at the Institute for Advanced Study, in Princeton, of which he was then the director. By that time, he was the epitome of Yeats's "smiling public man"—an extraordinary presence, combing a sort of fashionable elegance with a love of the outdoors. He was no longer doing active original research in physics, but he seemed to understand any idea instantly. The Institute's weekly seminars in theoretical physics were by far the most remarkable I have ever attended. There was an air of excitement and tension about them, partly because of the quality of what was being discussed but largely because of Oppenheimer's reaction to it, which could vary from extreme enthusiasm to the most caustic—almost cruel—kind of criticism I have ever heard. Having had this experience, I was, of course, eager to find out from Rabi how the young Oppenheimer appeared to his contemporaries. He was twenty-four when Rabi first met him. Rabi said, "He was very intense. He worked hard, but tried to give the appearance of not working. He was interested in everything. Pauli once said that for Oppenheimer physics was an avocation—that his real interest was psychoanalysis. I think he had been through analysis about that time. He was legendary, really."

I asked Rabi if he had found Oppenheimer easy to get along with, and he answered, "I found him excellent. We got along very well. We were friends until his last day. I enjoyed the things about him that some people disliked. It's true that you carried on a charade with him. He lived a charade, and you went along with it. It was fine—matching wits and so on. Oppenheimer was great fun, and I took him for what he was. I understood his problem."

"What was his problem?" I asked.

"Identity," Rabi answered. "He reminded me very much of a boyhood friend about whom someone said that he couldn't make up his mind whether to be president of the B'nai B'rith or the Knights of Columbus. Perhaps he really wanted to be both, simultaneously. Oppenheimer wanted every experience. In that sense, he never focused. My own feeling is that if he had studied the Talmud and Hebrew, rather than Sanskrit, he would have been a much greater physicist. I never ran into anyone who was brighter than he was. But to be more original and profound I think you have to be more focused."

In 1929, Pauli moved to Zurich, and both Rabi and Oppenheimer went along to work with him. Rabi's fellowship money had just about run out, and it appeared to him that his career in physics had almost come to an end. "I had no hope of getting a job," he told me. "It seemed to me that it would be very difficult to get one. Anti-Semitism was unbelievably rife in the universities and elsewhere. Because I really had no hope of getting a job, I didn't even write anywhere. I thought I would go back to New York and see what would happen and maybe, if worse came to worst, join my father in real estate. I didn't feel terrible about it; I had wanted to know physics, and I had had a tremendous experience. I thought, O.K., if you can do better, fine—if not, not. In the midst of all of this, a cable arrived—a radiogram from Columbia—offering me a job at $3,000 a year. This was to me, then, beyond the dreams of avarice." It turned out that Heisenberg, who had gone to America on a lecture tour, had heard that Columbia was looking for a theoretical physicist, and, without telling him, had recommended Rabi for the job. It came just in time. "The Depression began right afterward, and there were no jobs at all—for rich, poor, Jew, or gentile," Rabi said.

There is little doubt that the United States is at present the preeminent scientific and technological power in the world. To those of us who have grown up since the Second World War, this situation has a quality of something like inevitability; it seems that things could not and never will be any other way. In reality, this preeminence is a relatively new phenomenon. In the late 1920s, when Rabi was studying in Europe, American science, and particularly American physics, was generally held in contempt. In 1927, when Rabi arrived in Hamburg, he went to the library to look for *The Physical Review.* To his surprise, he discovered that the journal was held in such low esteem that instead of subscribing to it issue by issue the library—for the sake of completeness and to save postage—would order the entire year's worth of issues at the end of the year, and in due course they would be unpacked and put on a shelf. Having them there on a more current basis was felt to be unimportant in maintaining contact with the latest discoveries in physics, since these were being made largely in Europe and published in journals like *Nature* and the *Proceedings of the Royal Society,* in Britain, and the *Zeitschrift für Physik* and *Annalen der Physik,* in Germany. At that time, in fact, Rabi himself published the results of his work in these journals. Furthermore, little serious industrial research was being done in this country. American corporations—like many foreign corporations now, especially the Japanese—found that it was more economical to buy patents from abroad than to do research. "There were only a few laboratories in industry," Rabi recalls. "The Bell Laboratories, which were just beginning, and then General Electric were of importance for physics. They were very good. But for a country of this size—hardly anything. From the viewpoint of 1930, it is astonishing that by 1940 we had enough first-rate scientists to man the laboratories for microwave radar, atomic energy, and all the rest." A common fallacy is that this change was brought about by the arrival in this country of European scientists seeking refuge from the Nazis. In the first place, most of the great refugee scientists, like Enrico Fermi, arrived here late in the decade. (Fermi came to Columbia from Italy in 1939.) They ultimately had an important impact on the development of the atomic bomb, which began in earnest in 1942, but almost none of the development of radar, which, building on the fundamental earlier British work, took place at the Radiation Laboratory in Cambridge, Massachusetts, starting in 1940. This laboratory was staffed at the top mainly by American physicists, like Rabi and his contemporaries, and by their students and the students of their students.

The transformation in American physics did not come about by accident. It came about, in large measure, because a small but influential group of young American scientists—in physics, people like E. U. Condon, Oppenheimer, and Rabi—came back to this country in the late 1920s determined to make American science respectable. Rabi told me that one effect of his German experience was to "feel the greatness of America," and he went on, "It's true that America was backward in physics—really underdeveloped—but Condon and I and some others promised ourselves that we would end this. And we did. I told you the story about *The Physical Review* in 1927. Well, ten years later, it was the leading physics journal in the world."

I asked Rabi what, exactly, had given him this feeling about America when he was in Germany. "In the first place, I realized that my own education had not really been defective," he replied. "I was better prepared to read the literature and do something with it than most of the Germans at my level. It's true that there were some very great men there—no question about it. But the others? When you looked around for the second tier, you found that it just didn't exist. The general contempt felt for physics in the United States intellectually was a bit hard to take. A quotation from Arnold Sommerfeld at that time was typical. Albrecht Unsöld, the astrophysicist, who was a student of Sommerfeld's, had a fellowship to go to Berkeley. He was worried and fussy. Where would he live? What would he wear? What would he eat? How should he behave? And Sommerfeld said to him, *'Nehmen sie das nicht so ernst. Das Leben in Amerika ist gar nicht so schwer. Dort kann jeder junge Mann* assistant professor *werden.'* ['Don't take all this so seriously. Life in America is not so difficult. There any young man can become an assistant professor.'] The Germans' misunderstanding of the United States was so great that I was known as a chauvinist because I would argue with them all the time. There were plenty of things in the United States I didn't like—plenty of things. But not what they talked about. They didn't like the things about us that were good. After all, we had an honest-to-goodness democratic system—you could live in it."

The lack of a second tier, and of a genuine democracy, in German science contributed to the Germans' failure, a decade or so later, to make an atomic bomb and to their delay in developing radar until fairly late in the war. The situation in regard to the atomic bomb seems to be clear-cut. There were, of course, serious industrial difficulties, since so much of the country was under heavy bombardment. But those did not stop the development of the V-2 rocket, which was certainly a monumental technological feat. The main reason for the Germans' failure to produce a bomb was—despite what some German physicists have since said about their "humanitarian" motives—the fact that the great German nuclear physicists, such as Werner Heisenberg and Walter Gerlach, were never able to complete the first step: developing a nuclear reactor. Furthermore, they never realized how a reactor was best to be used in making an atomic bomb; namely, for producing the element plutonium, which would then be used in its pure form as the nuclear-explosive material, as in the bomb that leveled Nagasaki. At the time the atomic bomb was exploded over Hiroshima, the key German physicists had been rounded up by the Allied forces, shipped out of Germany, and interned in an estate about fifty miles from London. Their entire conversation after they learned of the Hiroshima bomb was monitored, and it was a full day after the news of the explosion before they realized that our "atomic bomb" was not an exploding reactor and the reactor had been made to use plutonium. (This is documented in the extraordinary book *Alsos,* by Samuel Goudsmit, the Dutch-born American physicist who headed the United States mission to determine the extent of Germany's atomic-bomb development.) Once the initial mistake had been made, no one in the German bomb project, apparently, had the authority to contradict Heisenberg, or even to argue with him. Rabi wrote in his book *Science: The Center of Culture,* published in 1970:

> A contributory cause of failure was the arrogance of German scientists. Thus Heisenberg, a scientist who, according to Dr. Goudsmit, ranks with Einstein and Bohr, was the head of a small inner circle, which was not receptive to ideas from lesser lights. Thus, when Fritz Houtermans came close to the idea of plutonium as early as 1941, no one took notice of his report because he was not in the favored circle. The failure to recognize the possibilities of plutonium was one of Heisenberg's—and Germany's—biggest mistakes in the atom bomb race. Professor Goudsmit charges the Germans with indulging in hero worship of Heisenberg and other top scientists to the extent that they refrained from thinking critically about their work. The uranium problem, like many others, he states, is too extensive to be grasped by one individual. A clash of ideas from many minds often is the key to success.

I asked Rabi if he had seen any intimations of the coming of the Nazis in Germany in the late 1920s. "Just the beginning," he said. "I hadn't heard of them. I first ran into them in Nuremberg. I didn't know what the hell it was about. Then, in Hamburg, you could see them in their *Fackelzüge*—torchlight parades. The Communists also had those, but it was clear that the Nazis were better organized. The German professors thought Nazism was just a joke, because the German spoken by the Nazis was so bad. However, Helen attended the Kunstgewerbe Schule, a school of applied art, and among the students there were quite a number of Nazis—very strong. Since she doesn't look particularly Jewish, they talked to her frankly about what they expected to do. They said the schools around there, and a beautiful public swimming pool, had all been built with American money from the Dawes loan." Charles Gates Dawes, an American diplomat, developed the Dawes plan for managing the German reparations payments after the First World War. " 'We are not going to pay that back,' the Nazis told Helen. And they told her how the next war was going to go. They were already training children to live and fight in the desert. There was no question about their fantastic anti-Semitism. At the time, though, I didn't know anything about it. I never had any problem in Germany because of being Jewish."

In fact, Rabi—as far as he knows—was the first Jewish member of the Columbia Physics Department. He started as a lecturer in 1929. In short order, he began to move up the ladder. "After the first year, even though I didn't publish anything, I was given a promotion to assistant professor," he said. "After the second year, I was given a raise. The third year, I still hadn't published much of anything, and they wanted to make me an associate professor." This hardly conformed to the usual image of "publish or perish" in major universities, so I asked Rabi what he thought his superiors had seen in him then. "I was the life of the place," he answered. "Students were flocking around, and I was in correspondence with and close to other physicists who were well-known, and so on. I was in the mainstream. G. B. Pegram, the Dean of the Columbia School of Mines, Engineering, and Chemistry, had great faith in me. When I joined the Columbia faculty, I taught two hours a week. There were full professors who were teaching fourteen. But Pegram was the boss, and no one dared to protest."

In time, Rabi taught most of the modern theoretical courses. He also ran a weekly theoretical seminar with Gregory Breit, a distinguished physicist who was then at New York University. (He later moved on to Yale, from which he retired a few years ago.) At that time, this was the only theoretical-physics seminar in New York. There are now about a dozen, in various physics departments in the city. In the beginning, Rabi had decided to return to pure theory. "I had some ideas—good ideas, actually—about solid-state properties, but they bored the hell out of me," he told me. "So I said, 'I'll use what I learned and do some molecular-beam experiments.' " This was at the beginning of the

Depression, and there was considerable doubt whether the universities were going to be able to continue to pay faculty salaries, let alone support research. "I remember going to the dean of the Graduate School in 1932 or 1933 and telling him that I would rather have my salary cut than take a cut in my research appropriation," Rabi said. At about that time, however, he had a fantastic piece of good fortune. Harold Urey, who was a professor at Columbia, and who won the Nobel Prize for chemistry in 1934 for his discovery of the deuterium isotope of hydrogen, received an offer of support for his research from the Carnegie Institution, in Washington. The Institution was willing to give Urey $7,600—a fortune at that time. "Well, Urey did one of the most extraordinary things imaginable," Rabi told me. "He gave me half of it. I had had nothing to do with his discovery. What a greatness in Harold Urey—what a tremendous magnanimity to do something like that! He had a deep faith in me. When he came back from receiving his Nobel Prize, he told somebody, referring to me, 'That man is going to win the Nobel Prize.' I don't know what he saw in me. I didn't think so—I didn't agree at all. But what a tremendous magnanimity! This money set me free. It made me independent of the Physics Department. Before that, I had to get permission to get started. I had to explain my research ideas to Dean Pegram. He was a wonderful man—brilliant. But after I would explain my ideas, he also had ideas, and it became a joint thing. I wasn't interested in that. I wanted to do my own thing. Once I had this money, this nest egg, I could start, and then if I needed money it was to fund an ongoing project. It was the greatest thing that could have happened to me. And what a greatness in Harold Urey!"

Having his nest egg, Rabi was not going to squander it. "I spent very little at a time, because, going around New York, you could buy 'junk' at a good price. When we needed a high-current generator, we bought it secondhand. We got vacuum pumps for eight dollars—not as good as the modern pumps, but good enough for our purposes. We salvaged a bunch of junk from a dismantled Poulsen-arc station in the Pacific. It was a copper mine for us. We did have a machine shop in pretty good condition. But we made the money last. The Carnegie people kept writing me letters saying, 'You're not spending money. It's bollixing up our books.' And I said, 'We're trying to be careful with it.' I was not about to give away my independence." So began the series of experiments that led, ten years later, to Rabi's Nobel Prize.

Rabi told me that of all the great physicists he met in Europe in the late 1920s, Niels Bohr was by far the most impressive. He also told me that one reason he had been pleased at being "thrown out" of Bohr's Institute in Copenhagen was that that way he had been able to escape the so-called *Kopenhagener Geist der Quantentheorie*. The *Kopenhagener Geist* is the now widely accepted interpretation of the limitations of measurement in the quantum theory. It was formulated by Bohr and nurtured over the years by him, and it is summarized in the Heisenberg uncertainty relations, which were inspired by Bohr's analysis. These relations emerge automatically in the Heisenberg-Schrödinger quantum-mechanical formulations. The key element of the *Geist* is the necessity to renounce asking certain kinds of questions about physical events—above all, on the atomic scale—since, while these questions may seem plausible, it turns out that when they are analyzed they involve measurement procedures impossible in principle to carry out. For example, if one has a collection of radioactive nuclei, one might like to ask when a given nucleus—say, the one in the corner of the box—will decay. To this question, the theory has no answer; in fact, the theory denies that it is a properly formulated question. The theory tells you only the probable length of time—the "lifetime"—that the particle will "live": the length of life of an *average* particle. According to the theory, there is some small probability that any given particle in the sample will live a much longer—or much shorter—time than the average. The theory deals in likelihoods and probabilities. One cannot predict when the given nucleus will decay, for to do that one would have to "open up" the nucleus to look inside, and such a measurement would require enough energy to break up and destroy the nucleus. This sort of argument is typical of the *Kopenhagener Geist* arguments.

When Rabi told me he didn't believe in the *Kopenhagener Geist,* I was somewhat surprised, since I had always thought that, having been around at its creation, so to speak, he must have been indoctrinated in it. "I am very skeptical of it, because I have not found that point of view to be useful except as an easy way to talk to non-physicists," Rabi said. "But Bohr was such a forceful and charming person that you couldn't help being overwhelmed by his point of view, whether you agreed with it or not. Ultimately, you had to succumb. He had enormous strength and personality. But, basically, I was not completely persuaded, and I am not to this day completely persuaded that this interpretation is what will remain. As time goes by, its power becomes greater and greater, and it has not been in any way demonstrated to be wrong. Still, I don't think it's right enough to suit me aesthetically. I don't know how to ask a question and have quantum mechanics tell me how to set up a piece of apparatus that corresponds to the asking of that particular question. If you try to do the simplest thing using quantum mechanics—using the idea of a particle, say—it becomes vastly complicated. It relates observations, but it doesn't tell you how to make an observation. In this sense, for me it lacks immediacy. But the power of the quantum theory within its limitations is simply unbelievable. It was certainly unbelievable to me at the beginning. For years and years, I presented it with situations to see where it would fail. I thought, 'It's got to break somewhere.' But it didn't."

Rabi went on, "But the questions you can ask are limited. Things I can do with ease, classically, like angular momentum, I can hardly do at all quantum-mechanically. On the other hand, there's the miracle of doing something like making an electron-positron pair! One actually creates a remarkable thing like an electron. It's a marvelous thing. I don't see how it's made. It just appears. It's a kind of materialization—the ghost shows up in reality. There it is. You can calculate how many electrons will be produced, and with what probability. But how was it born? What was it made of? It's this kind of question that, as an experimenter, I would like to see answered. The theory doesn't answer the sort of question that led me into physics in the first place. I wanted to know what the thing really was. The theory leaves you still baffled if you have been raised in classical mechanics. I wasn't raised very much in classical mechanics, so I'm not insisting on classical mechanics. As an experimenter, I'm just not satisfied with trying to find out what is going on and having only this very mathematical or verbal kind of answer. You know that, according to quantum theory, if two particles collide with enough energy you can, in principle, with an infinitesimal probability, produce two grand pianos. How would they have been made?"

I asked Rabi if he had ever discussed any of this with Bohr. "I couldn't," Rabi said. "This work was his life. There was no point in trying to tell him that I thought it was irrelevant to the sort of things that an experimenter actually does in the laboratory. I felt that in this he was very profound about things that don't really matter. But one was not going to tell him that."

From the early 1930s on, the theme of Rabi's experiments was, as he put it, "to play the changes" on the original Stern-Gerlach setup. On the one hand, he was trying repeatedly to push the quantum theory into a corner, and, on the other hand, he was trying to make high-precision experiments dealing with magnetic moments and what are known as nuclear spins.

The concept of spin came into physics in 1925, when two then extremely young Dutch graduate students—Samuel Goudsmit, the author of *Alsos,* and George Uhlenbeck, who is now a professor emeritus of The Rockefeller University—suggested it as an additional source of magnetic energy, in order to patch up certain discrepancies that had arisen in atomic physics. (It had been independently invented about the same time by Rabi's former fellow student Ralph Kronig. He was talked out of publishing it by, among others, Pauli. Hence he never got full credit for this work. This story is often cited by physicists as an example that it is sometimes best not to take the advice of senior people about publishing an idea that may seem a little crazy if the idea accounts for a number of experimental facts.) A revolving electron acts like a current in a wire in that it produces an associated magnetic field. Now it turns out that the atom has an extra source of magnetism, which is present even if the electron is not revolving. This "intrinsic" source of magnetism is related to the so-called spin, which is actually a misnomer, for although in the simpleminded picture of spin the electron is supposed to be spinning on its own axis, like the earth, a straightforward calculation shows that the surface of such an electron would be moving faster than the velocity of light—something that according to the theory of relativity is impossible. Spin is one of those numerous instances described by Rabi in which the quantum theory defies intuitive visualization. There are rules that tell how to work with spin, and, as far as the theory is concerned, the concept and the rules are the same. The rules for the spin are that it acts physically and mathematically like a quantum-mechanical angular momentum. But this angular momentum is present even when the electron is at rest. Hence, in general, there are two sources for the angular momentum of an electron—one is its orbital motion and the other is its spin. But when Goudsmit and Uhlenbeck invented the spin, there was considerable resistance to the idea, since at that time people were not accustomed to forgoing intuitive visualization. Bohr accepted the notion of spin almost at once, but Pauli did not, and on one celebrated occasion, when Bohr was traveling from Copenhagen to Leiden, Pauli went to the Hamburg railroad station, where Bohr's train had a brief stopover, to try to warn him against the spin. A little later, Pauli changed his mind, and he invented the basic mathematical apparatus by which spin is now described in the quantum theory. A "spin magnetic moment" of an electron, say—despite its romantic sound—is a measure of the strength of the electron's magnetic field, and is of the same nature as, for example, the magnetic moment of the magnetic field in a compass, but vastly smaller.

Not long after Goudsmit and Uhlenbeck made their hypothesis for the electron, it was suggested, in order to explain some results in atomic spectra, that atomic nuclei must also have spins and magnetic moments—something that could be attributed largely to the spins of neutrons and protons, of which the nuclei are composed. In 1933, Stern succeeded in making a quantitative measurement of the magnetic moment of the proton, which turned out to be about three times as large as he had expected. He had expected one nuclear magneton, but his experiment found 2.8. His result not only demonstrated the existence of a nuclear magnetic moment but also showed that the magnetic character of the proton was much more complicated than could be accounted for by the standard theory of the electron. This "extra" magnetic moment is now known to result from the nuclear forces that bind the neutrons and protons together. The forces come from the exchange of charged mesons—particles with masses heavier than those of the electrons but lighter than those of the protons and neutrons. But although theorists feel certain that the mesons exchanged between neutrons and protons are responsible for the extra magnetic moment, it has turned out to be exceedingly difficult to produce a theory that gives the precise magnitudes.

In 1931, Breit and Rabi produced a now classic theoretical formula that gives the response of a nucleus with arbitrary spin surrounded by a cloud of electrons with their spins—making up a neutral atom—to a non-uniform magnetic field. If, for example, the total spin of these electrons is, in units of $h/2\pi$, simply $1/2$, then a beam of such atoms placed in a strong non-uniform magnetic field will be split in two, as in the original Stern-Gerlach experiment. However, as the field is made weaker, the electron spins and the nuclear spin act jointly—are coupled—and the number of beams into which the original beam splits becomes much larger. To take a specific example, the sodium nucleus has a spin of 3/2, the sodium electrons a spin of 1/2. According to quantum theory, a spin of 1/2 and 3/2 can combine to give only two possible total spins: 2=3/2+1/2, or 1=3/2—1/2. If one studies the Breit-Rabi formula, one finds that in a weaker field this means that the original sodium beam will be split into eight sub-beams, corresponding to the various possible configurations of the two spins. (There are five for the spin two configuration and three for the spin one.) To test this, and to measure the nuclear spin of sodium, Rabi devised an experiment, which he carried out in 1933 with the late Victor W. Cohen—the son of Morris Cohen, the philosopher at City College—who worked at Brookhaven National Laboratory in recent years. They separated a beam of sodium atoms in a strong magnetic field and, using a system of slits, selected a part of the beam whose atoms had uniform velocities. They ran this part into a weaker, non-uniform magnetic field, which split the beam in four. In their original selection, they had chosen half of the possible magnetic states, so they produced only four of the eight possible beams. These four beams were next refocused with a third non-uniform magnetic field—a strong one—and were then detected by letting them impinge on a movable tungsten wire. Not only did the result—the demonstration that there *were* four beams—verify the Breit-Rabi formula but it also provided an extremely powerful new way of measuring the spins of nuclei. One could simply count the number of beams. Recently, Rabi recalled, "You could read the maxima where the four beams arrived. You could play with them. They were just great—a tremendous thrill. The experiments were very elegant and very definite. They displayed the principles of quantum mechanics, which, after all, at that time were not so ancient as they are now. I got a great deal of satisfaction out of realizing that I wasn't just horsing around with nothing. My ideal was at the end of the day to have an answer—a measurement of a nuclear spin."

In 1934, with two other young collaborators—Jerrold Zacharias, who is now professor emeritus at MIT, and J. M. B. Kellogg, who is now retired from Los Alamos—the growing molecular-beam group measured the magnetic moments of the proton and of the nucleus (the deuteron) of the heavy isotope of hydrogen, which had just been discovered by Urey. The proton moment was in agreement with Stern's earlier result. The molecular-beam group—which also included, among others, Sidney Millman, who has just retired from Bell Laboratories; Polykarp Kusch, now at the University of Texas at Dallas; and Norman Ramsey, now a professor at Harvard—was not a factory of physics papers; it published only three or four a year, but each one was a jewel.

In 1937, Rabi devised a method for magnetic-moment measurements that became the central technique in all modern molecular- and atomic-beam experiments, and has transformed the subject into a veritable industry of experimental physics. As before, there are strong non-uniform entry and exit magnets to refocus the beam on a detector. However, in the weak uniform magnetic field between those two magnets Rabi put an oscillator, which produced an additional weak magnetic field that oscillated periodically—like the sending station of a radio transmitter. Rabi arranged things so that he could control the frequency of this oscillation with extreme accuracy. The arrangement was almost like a tuning fork with a variable pitch. When, say, a sodium atom entered the weak-field region, it was in one of the eight possible states of angular momentum. Now, by tuning his oscillator, Rabi could cause the atom to make a quantum jump from one such state to another. At a certain frequency, he was feeding in just the right amount of energy to cause the jump. Whenever this happened, the atom was thrown out of alignment and that beam stopped focusing on the detector. Thus, by determining the frequency at which the beam stopped focusing, Rabi was making a direct measurement of the energy required to cause the spin to jump, and this energy, it turned out, was proportional to the magnetic moment. Hence he had a technique for measuring magnetic moments, and this technique, with variants, has turned out to be incredibly accurate. To get some idea of how accurate these methods are, one may note that the pre-oscillatory-field techniques produced results valid to two significant figures. The first Rabi experiments produced results to three significant figures, and, just after the Second World War, Polykarp Kusch measured by resonance methods (the usual term for oscillatory-field techniques, because one is allowing the frequency of the field to "resonate" with the energy needed to make the quantum jump) the magnetic moment of the electron to six significant figures. (He shared the 1955 Nobel Prize in physics for this work.) It is now known to nine significant figures, or three parts in a billion.

The students and young faculty members who worked with Rabi at this time were not only from Columbia but from elsewhere in New York City. This was during the Depression, when the smaller institutions simply could not afford to operate their own physics laboratories, so many young physicists had to come to Columbia to do their research. Zacharias, for example, was an assistant professor at Hunter College, teaching sixteen hours a week. In addition, he worked about fifty hours a week in Rabi's laboratory. "It was the happiest period of our lives," Rabi told me. "We liked to work together, and we worked every day of the week. During the year, we would take something like three or four days off. The phenomena were so beautiful—really so beautiful—that people worked terribly hard and very joyfully." At the height of the activity, Rabi had about fifteen people working in his laboratory—a large number for that era in physics—as well as half the department's students and half its money for research, which meant that his lab got $8,500 a year.

Besides Kusch, two of Rabi's other young associates of that period went on to win the Nobel Prize. Willis E. Lamb, who joined Rabi's group as an instructor in 1938, shared the prize with Kusch in 1955; Lamb's experiment had involved the measurement—again, a variant of the molecular-beam resonance technique—of the so-called Lamb shift in the energy levels of hydrogen, the shift being a tiny displacement of the energies of the electron circling the hydrogen nucleus that arises as a subtle effect in the quantum theory of radiation. Kusch's and Lamb's discoveries were the inspiration for an entirely new development in the quantum theory of radiation, and one of the prime movers in this work was Julian Schwinger, a Rabi student who shared the Nobel Prize with Richard Feynman and Shinichiro Tomonaga in 1965 for this theoretical development. Rabi's discovery of Schwinger, who was one of the most influential teachers of my own generation of physicists (I took many courses from him when I was at Harvard in the 1950s), was remarkable. "It was sort of romantic, in a way," Rabi recalls. "You can fix it to a year—1935. Einstein, Boris Podolsky, and Nathan Rosen had just published their famous paper"—on an apparent paradox in the quantum theory of measurement. "I was reading the paper, and my way of reading a paper was to bring in a student and explain it to him. In this case, the student was Lloyd Motz, who's now a professor of astronomy at Columbia. We were arguing about something, and after a while Motz said there was someone waiting outside the office, and asked if he could bring him in. He brought in this kid." Schwinger was then sixteen. "So I told him to sit down someplace, and he sat down. Motz and I were arguing, and this kid pipes up and settles the argument by the use of the completeness theorem"—an important mathematical theorem used frequently in the quantum theory. "And I said, 'Who the hell is this?' Well, it turned out that he was a sophomore at City College, and he was doing very badly—flunking his courses, not in physics, but doing very badly. I talked to him for a while and was deeply impressed. He had already written a paper on quantum electrodynamics. So I asked him if he wanted to transfer, and he said yes. He gave me a transcript, and I looked at it. He was failing—English, and just about everything else. He spoke well. I said, 'What's the matter with you? You're flunking English. You speak well, and you sound like an educated person.' He said, 'I have no time to do the themes.' I tried to get him admitted to Columbia on a scholarship. I saw the director of admissions. He looked at the transcript and said, 'A scholarship?' He wouldn't even admit him. Well, Hans Bethe was passing through, and I asked him if he would read Schwinger's paper. He read it and thought well of it. So I asked Bethe to write me a letter. He did. And, armed with this letter, I got Schwinger admitted. He entered Columbia as a junior and actually made Phi Beta Kappa. He turned over a new leaf." (George Uhlenbeck also remembers being asked by Rabi to plead Schwinger's case.)

Rabi went on to tell me about Schwinger's graduation. "One Sunday morning, the dean calls me up and says, 'Schwinger has all his course requirements to graduate, but he doesn't have enough maturity credits.' " At that time, Columbia gave grades for performance in the course work and also something called a maturity credit, which was a measure of the degree of difficulty of the course. It was an attempt to stop students who had taken nothing but snap courses from getting a degree. "I was sore that the dean had called me on this thing. So I said, 'I suppose if Schwinger doesn't have enough maturity credits, he can't graduate.' Silence. Then the dean said, 'I'll be damned if I am going to stop Schwinger from graduating just because he doesn't have enough maturity credits.' That's what I wanted to hear. I didn't want to beg the dean to graduate Schwinger, but if he hadn't he would have been a damn fool." At the time of his graduation, Schwinger had just finished completing the material for his Ph.D. thesis, and he remained at Columbia two years longer, to complete his doctoral residence requirement. He then went on to work with Oppenheimer in California, and has been one of the foremost American theoretical physicists ever since.

Physics has now become an extraordinarily complex professional discipline, broken up into dozens of specialties and subspecialties, some involving experiment and some involving theory. No one today can possibly have a complete grasp of every one of them. Above all, there is a primary differentiation between experimental and theoretical physicists. In most universities, the two groups receive rather different educations after the introductory courses. This arrangement runs the risk of producing what Rabi once referred to as "people who call themselves theorists who can't tie their shoes and people who call themselves experimenters who can hardly add." When Rabi took his degree, such a situation really did not exist—at least, at Columbia—in large part because there was very little instruction there, at that time, in modern theoretical physics. The sharp division, according to Rabi, developed in Germany at the beginning of the century. He told me that if he had been a German and Stern had asked him to do an experiment, he probably would not have done it, since he was otherwise doing theoretical physics. Rabi believes in giving all physics students the same kind of education, "by never telling the student which he is." He explains, "You give the student the fundamentals, and then he comes around to look for someone to work with on a dissertation. He decides with whom he would like to work. If he wants to work with Mr. X, and Mr. X is a theorist and takes him on, then theoretical physics is what he will do." In consequence of this educational philosophy, the Columbia theoretical physicists have always been very closely associated with experimental developments, and the experimenters have had a thorough grasp of theory. Rabi believes that experimenters should generate their own experimental ideas, and not depend on theoretical physicists to tell them what to do. "The last person a young experimenter should ask, in my belief, about an experimental program is a theoretical physicist," he declared. "Not that theoretical physicists are stupid, but they have their ideas, and they want to know the answer to their own problems in their terms. We would not have advanced very far in basic discoveries with a concentration on theory alone. We don't teach our students enough of the intellectual content of experiments—their novelty and their capacity for opening new fields. If you follow a theorist, you come in and say, 'Now what shall I do?' And then you do it, and come back and say, 'What have I done?' I don't know why people work that way. My own view is that you take these things personally. You do an experiment because your own philosophy makes you want to know the result. It's too hard, and life is too short, to spend your time doing something because someone else has said it's important. You must feel the thing yourself—feel that it will change your outlook and your way of life. You must bring it back to the human condition, the human expression—much closer to what the artist is supposed to feel."

I suggested that perhaps this separation of specialties was necessary because of the cost of doing experiments, not only in physics, but in all branches of modern science. This idea didn't get very far with Rabi. "I don't know why people are so impressed by money," he said. "I don't know why. I think that the first experiment that is done pays for the machine. After that, it's gravy. But if you sell your soul to a machine it's terrible. I used to feel that way about my apparatus. There it was. It had to be cleaned. It had to be kept leakproof, and so on. And then I would say, 'Now, who's master here—this machine or me?' I don't think we should worry so much about money. Just because a machine costs a lot of money doesn't mean you have to use it for a lot of dumb experiments."

In 1938, by which time, as Rabi put it, "the molecular beams were going like a house afire," he took a sabbatical leave, during which he did research and consulted with colleagues both at the Institute for Advanced Study and at Columbia. "It was the wrong time for a sabbatical," Rabi told me. "The stuff was going great here, and I had to commute to New York almost every day." Einstein was at Princeton by then, and Rabi had a chance to spend some time with him. "There was one evening after a concert when we all went back to his house," he told me. "Einstein brought out a bottle of lovely cognac. He didn't drink. His secretary didn't drink. Nobody drank except me. So I drank the cognac, and the more I drank, the mellower I got. But then everybody else did, too. It was very interesting. At one point, Helen Dukas, Einstein's secretary, brought out Einstein's report cards. It turned out that he had been a pretty good student." Einstein, like most of the refugee scientists, had become more and more alarmed by the rise of Hitler, and feared that the United States would stand by and allow Hitler to take over Europe. By this time, Rabi was thinking about what he could do, as a physicist, to help in what he was sure would be a war in which America would become involved. "I really felt I ought to do something," he said. "And later, with the fall of France, I became desperate to get into the war."

The war for Rabi meant, primarily, radar: the practical development of modern radar. Science-fiction buffs are fond of claiming that one of their own—Hugo Gernsback—"invented" radar. It is true that in 1911 Gernsback began a series of science-fiction stories which ultimately resulted in an all but unreadable novel, called *Ralph 124C 41 +*, about an American scientist living in 2660 who invents something that we would now call radar. This, as it happens, is one instance in which Gernsback, whose track record for technological prescience was exceedingly strong, was several years behind reality. The idea that electromagnetic radiation is generated by oscillating electric charges originated with James Clerk Maxwell in the late 1860s and was demonstrated experimentally by the German physicist Heinrich Hertz in 1888. The basic mechanism of this radiation is rather simple to characterize in terms of electrons. When an electron is set in an oscillatory motion through the action of an electromagnetic field, the electron radiates—or, in modern terms, when an electron is accelerated, it gives off photons, which are quanta of radiation. In a conductor—a metal, say—the electrons are free to move about, and when an electromagnetic wave, which is a collection of quanta, impinges on such a surface, the electrons are set in motion. These electrons become transmitters of radiation. Thus, the surface gives off radiation; that is, it reflects some of the radiation that has impinged on it from the outside. Some of this radiation is returned toward the source that generated it, and the returning radiation pulses are the echo from the conducting surface they have encountered. (This phenomenon was also first demonstrated in the laboratory by Hertz.) As early as 1900, Nikola Tesla, writing in *Century Magazine,* pointed out the utility of radio echoes in finding moving objects—such as ships. Then, in 1904, a German engineer, Christian Hülsmeyer, took out patents in several countries for a radio-echo device to help prevent ship collisions. But, presumably because collisions were relatively infrequent at that time, little was done to develop the device.

In the mid 1920s, two Americans, Gregory Breit and Merle A. Tuve, conceived the notion of "pulse ranging"—sending out radio waves in short bursts, with a relatively long interval between. (In a modern radar, these pulses last for something like a millionth of a second.) The virtue of pulse ranging is that one can time the delay between the sending of such a pulse and the receiving of its echo with great accuracy. (The radar transmitter is turned off between pulses, leaving the set free to pick up echoes.) The delay indicates the distance of the target, because the velocity of light is known to a high degree of accuracy, and radio waves, like all electromagnetic radiation, move with the speed of light. Thus, the delay can simply be multiplied by the speed of light, and the result is the round-trip distance. The word *radar,* which was coined by the United States Navy (it is an acronym for "radio detecting and ranging," and even the name itself was a military secret until 1942), almost always refers to pulsed radiation. The first application of the radar method was in measuring the height of the ionosphere, a band of ions and free electrons extending from a height of fifty miles above sea level to a height of two hundred miles above sea level, with two regions of maximum density. One extends from fifty miles above sea level to seventy-five miles, and the other from one hundred miles above sea level to two hundred, with some variations caused by changes in atmospheric heating—night and day, summer and winter. It is the existence of the ionosphere that makes radio transmission possible beyond the visible horizon, since radio waves can be bounced off the charged particles of the ionosphere. In this respect, the ionosphere acts as a conducting surface.

In the 1930s, both the United States and the European countries began working on radar for military purposes. The problems that its development posed were two-fold—the need for power and the need for high directional accuracy. The radar pulses must be focused by the transmitting antennas into a narrow beam to pinpoint the target. It is a basic principle of physics that the longer the wavelength being transmitted, the larger the antenna that is needed to focus the beam sharply. In radar, a fifty-centimeter wavelength is considered fairly long, and this is the sort that was utilized in the early radars. Hence the original radars needed gigantic antenna towers totally unsuited for installation in ships or planes. Such radars, with relatively long wavelengths, were employed by the Germans in 1940 to direct anti-aircraft fire. The British, under the direction of Sir Robert Watson-Watt, head of the radio department of the National Physical Laboratory, had begun to set up radar warning systems on the coast as early as 1935. The real breakthrough in the art of radar began early in 1940, when two British researchers—J. T. Randall and H. A. H. Boot—invented the so-called multi-cavity magnetron, a device with many resonant cavities in which radio waves were generated that could produce a powerful pulsed beam at microwavelengths: ten centimeters and less. Thanks to the magnetron, highly focused and powerful pulsed beams could be produced by antennas the size of large salad bowls—light enough and small enough to be carried on planes and ships. It is probably no exaggeration to say that the air war against Germany and Japan was won because of the invention of the magnetron.

In September and October of 1940, the British sent a technical mission to the United States under the direction of Sir Henry Tizard, a scientist working with the Air Ministry. The British brought sample magnetrons and proposed that a laboratory be set up here for the development of airborne microwave radar. By this time, they had had something like five years of experience with the organization of a laboratory that was doing work for the military but was staffed by university physicists and engineers. This symbiosis of university scientists and the military, which had worked so successfully for the British, became the model of this country's own wartime military-scientific laboratories. By November, such a laboratory was begun here on a very small scale—the Radiation Laboratory at MIT, in Cambridge, under the leadership of Lee A. DuBridge, who was on leave from the University of Rochester. (After the war, DuBridge became president of the California Institute of Technology.) The laboratory began with about thirty people—mostly nuclear physicists from the universities. It existed for sixty-two months, at the end of which it had 4,000 people.

Rabi, who became associate director of the laboratory soon after it was founded, recalled how he had joined it. "I went to Lee and volunteered," he told me. "He was a bit reluctant, because there were suspicions about people from New York—Communist sympathies, and so on. But anyway, I persuaded him, and on November 6—the day after Election Day—I took off for Cambridge. A few people were there ahead of me. Oddly, I had already been cleared to receive highly classified material, but DuBridge hadn't. He was a great director—very well liked, and capable of motivating the members of the laboratory. The lab was created to develop airborne microwave radars that would help the British shoot down night-flying bombers. The job of developing the new equipment could be divided up into a number of specific fields—the transmitters, the receivers, the antennas, and so forth. I took the transmitters, which meant the magnetrons—the further development of the magnetrons. I'd never seen a magnetron before. In fact, I had never done any electronics. So I said, 'Well, one of these MIT people will tell me how it works.' I found out that they didn't know, either. There was at that time no good general theory of the magnetron. We started learning about it and the other components, and developing them further."

Rabi took advantage of a Sigma Xi lectureship, which allowed him to visit several universities, to recruit physicists. "We did as much as anyone to stop all research in the United States on non-war-related physics, by taking people out of the universities," Rabi said. "I didn't have any power, and didn't need it. I simply went around to the universities talking to people. I would go to some place and say, 'Now look, we have a laboratory in Cambridge. I can't really tell you what it's about. It isn't a very good laboratory, but—hell, I'd like to see you there in about two weeks.' And they'd come."

I remarked, "They must have had a lot of faith in you."

"They must have," Rabi answered. "For example, there was one man who was exceedingly anti-Roosevelt. He said to me, 'That man is going to get us into the war. You just mark my words; he's going to get us into the war.' I said, 'Yes, yes, let's not talk about it. You just come. We need you, and you show up.' And he did. Then when we did get into war, he came up to me and said, 'Didn't I tell you he'd get us into the war?' If anyone had been so frank some years later, he would never have been cleared. But this man came, and he did a very good job, and made his mark in the laboratory. We had a remarkable group of people, and if you were to read the roster you would think it was a nuclear laboratory, because so many of the people in it were from nuclear physics." George Uhlenbeck was the leader of the group in theoretical physics, and Julian Schwinger was recruited and became a legendary part of the Radiation Laboratory. He worked all night and slept during the day. "At five o'clock, when everybody was leaving, you'd see Schwinger coming in," Rabi said. I was once told that people would leave unsolved problems on their desks and blackboards, and find when they returned the next morning that Schwinger had solved them. "The problems he solved were just fantastic," Rabi continued. "He lectured twice a week on his current work. As soon as Schwinger would make an advance, guys all around—Dicke and Ed Purcell and so on—would invent things like mad. All sorts of things." R. H. Dicke is now a professor of physics at Princeton, and Edward M. Purcell a professor of physics at Harvard. Purcell shared the 1952 Nobel Prize in physics with the Swiss-American physicist Felix Bloch. Their experimental work on nuclear magnetism was along the lines begun by Rabi in the 1930s.

"In 1944, after the successful Allied landing in France, I felt that the war was essentially won and that our laboratory would have little more to contribute, because of the length of time—about two years—required to put newly developed equipment into the field," Rabi said. "At that point, only our laboratory and a very few others had real expertise in the field of radar in this country. I felt that our group should write a series of books on what we had learned, and that as the material became declassified these should be made available to the public. For example, if a congressional committee wanted to know what we did with the millions we spent, we could say, 'Here are the books. You can read them.' " The writing project was begun in the fall of 1944, under the direction of the late L. N. Ridenour, who, according to Rabi, had an enormous ability to organize writers and editors. More than 400 of the people from the laboratory worked on the books, and they produced twenty-eight volumes—all in one year. As Rabi pointed out, "these volumes became the basic texts in the fields of radar and electronics for the next two decades."

A cardinal element in the initial success of the wartime development of radar in the United States, according to Rabi, was the willingness of the people at the Radiation Laboratory to learn from the British. "Everything we knew at the beginning about radar we learned from them," he told me. "They were terrific. They really were tremendous, both technically and what I call philosophically. How to use it. What do you need and why do you need it and how would you use it when you got it? What are its limitations? What are its strengths? A great deal depended on how you used radar and what for. The British could do wonders with a few sets. We came to know a great deal, not because we were smarter, but because we were smart enough to listen to the British, who had had real operational experience."

One of the most important things the Americans learned from the British was that a group of university scientists could set up a pattern of cooperation with the military. The key to the whole set of successful relationships that eventually emerged during the war—for Rabi, at least—was an incident that came to Rabi's attention soon after the creation of the laboratory. Sir Mark Oliphant, a British nuclear physicist, who is now governor of South Australia, was in charge of a laboratory at the University of Birmingham, in England, which was making radar tubes for British military use. It was in this laboratory that the magnetron had been invented. A delegation from the military came to him one day and asked for a tube having certain characteristics. Oliphant asked what the tube was going to be used for, and was told that this was a highly classified matter and the military men were not permitted to say. Oliphant replied, "No information, no tube." And "No information, no tube" became Rabi's guideline in all his dealings with the military.

"We had to set relations with the military," Rabi said. "I was very forceful about that myself. For example, one group came from the Navy. They wanted certain black boxes, which they described, to be developed with certain voltages, and so on. I asked, 'What are they for?' Their exact answer was, 'We prefer to talk about this in our swivel chairs in Washington.' I didn't say anything. Neither did I do anything—except continue to develop three-centimeter radar. They came back some six months later—the same thing. I said, 'Now look, let's stop kidding. Bring your man who understands tactics, bring your man who understands radio, bring your man who understands aircraft, and we'll talk about your problems.' By that time, we had learned a lot about radar and tactical military applications. Well, the Navy did. We found that their problem was to knock off Japanese aircraft spying on ships. It turned out that they needed a shipborne height-finding radar to supplement and guide the radar equipment already in their carrier-based planes. And we made an agreement with the Navy. We'll develop that if you and we can do the whole thing together—a partnership. We're in this war together. We can talk about the whole thing, whatever it is, and then our side will do its best to develop the appropriate radar. Which it did. It was a fantastically great radar—a very effective thing. As time went on, we set up an effective pattern of interdependence with the military. Fortunately, our money did not come from the military directly but from another government agency, the Office of Scientific Research and Development, under Dr. Vannevar Bush. Actually, we would use this money to develop a particular radar. We would then try to interest either our military men or the British in it. If they were interested, then they financed the production. After we learned to get along with the military men, we grew to have a deep respect for them—respect for their devotion and hard work. We got along with them once they saw that we were not there to take anything away from them but actually to help them."

At the time the United States entered the war, radar was so new that it took a while for the various branches of the military to appreciate it. Rabi recalled, "We made a radar, my group—our first radar. It was an airborne three-centimeter radar. Then we made a ten-centimeter shipborne radar. The Navy put it on a destroyer that was stationed in London. The destroyer was to go out and measure things in connection with submarine detection. This radar was for location and detection—to detect things. It was a lash-up job. From that, they later made the real engineered and manufactured stuff. Well, I got the reports that the captain of the destroyer was sending in. There were repeated reports of failure—'Equipment broke down,' and 'Equipment broke down,' and so on. It was an honest job of reporting on a new piece of equipment and its reliability. You know how the Navy feels about reliability, and our radar wasn't perfectly reliable at that stage by any means. I saw that this captain could give us—give the whole project—a terrible black eye. So I went down and said, 'We'll have to make a few repairs. We'll take it back.' Once they lost that radar, they began to understand what it was. Because with that radar the ship could come in at night. Their wives could expect them for dinner. They could detect landmarks. They could navigate in fog. And so forth. They were like a man with a weak heart—in this case, the radar—who knows his trouble and his pain and has to learn to live with them. After a little while, we began getting requests to bring it back. I kept saying, 'We really have quite a lot of work to do on it.' Then they began to beg very hard, and we brought it back sort of improved. If I had gone down there and argued with them, we wouldn't have got anywhere. We would just have got them mad. After we returned it, the reports were excellent, and every once in a while they would ask us to help them fix it up. What had developed was a mutuality of interest."

After hearing this story and several others, I told Rabi that what impressed me about them was that he had shown a special instinct for dealing with people in extraordinary situations. "It's a matter of understanding a situation and adapting yourself to it," he said. "There were many mistakes I made in my life by being tactless and abrupt. But if there was something positive to be done, I really tried. I find it very hard to tolerate fools, and I can be quite rough. But if there is something to be done, and these are the people to do it with, what choice do you have?"

In the fall of 1942, J. Robert Oppenheimer, teaching alternately at the California Institute of Technology and at Berkeley, led Brigadier General Leslie Groves, a general in the War Department's Office of the Chief of Engineers, to Los Alamos, an isolated site in the mountains of New Mexico, not far from Santa Fe. Groves had the responsibility of recommending a director for the embryonic Manhattan Project, to make an atomic bomb. He was favorably impressed by Oppenheimer, who was then thirty-eight. Oppenheimer had spent a great deal of time in New Mexico, where he had a ranch. He had visited Los Alamos on horseback. From the beginning, he had been in on the discussions among the physicists about prospects for making a nuclear explosive, but, Rabi told me one day, "he was absolutely the most unlikely choice for a laboratory director imaginable." Rabi went on, "He was a very impractical sort of fellow. He walked around with scuffed shoes and a funny hat, and, more important, he didn't know anything about equipment. But he did a splendid job, a really remarkable, wonderful job. There were in my experience two outstanding wartime laboratory directors—our director, Lee DuBridge, and Oppenheimer. DuBridge was also a good administrator, and the lab ran very well. Oppenheimer's lab ran with flamboyance and excitement. It didn't run any better than ours, but it was an extraordinary experience. The whole thing lasted a little over two years—years that fixed the people who were there for life. It was their great moment."

Rabi went on, "I was in on the beginning of that, also. I was asked by Oppenheimer to join it and be the deputy director—take charge of the experimental work. I refused, because I was serious about the war. Maybe we could have won it without the atomic bomb, which was at that time a very problematical thing, but without radar we could have lost it. I did go to Los Alamos every once in a while, chiefly as a troubleshooter for Oppenheimer. He'd get into trouble with the Central Europeans and some of the other people. Oppenheimer had decisions to make which he wanted to talk over with a friend, and by that time I had had a great deal of experience with the problems of a laboratory and with problems with the military. Of course, Oppenheimer and I had known each other since our days with Pauli, and we had kept up our friendship. But I never went on the payroll at Los Alamos. I refused to. I wanted to have my lines of communication clear. I was not a member of any of their important committees, or anything of that sort, but just Oppenheimer's adviser."

Rabi was in the New Mexico desert, at Alamogordo, on July

16, 1945, when the first nuclear explosion in human history took place. In the only book of a popular nature he has ever written—*Science: The Center of Culture,* which was published in 1970—he describes the events of that morning. He writes:

> On July 16, 1945, I was out in the desert in New Mexico; not many yards away was Mr. James B. Conant, then president of Harvard University. There were other men there, also. We were awaiting the tests of the first large-scale release of atomic energy.
>
> In the desert there, the site chosen had been named "Journey of Death" hundreds of years ago. This site was chosen because it was supposed to be far away from human habitation. Nine miles away from where we were, there was a tower about one hundred feet high. On the top of that tower was a little shack about ten by ten. In that shack was a bomb. The whole experiment had been in preparation for almost a year. Hundreds of men took part in it. There were many rehearsals, and this particular morning it was to go off. At first, the announcer said: "Thirty seconds"—"Ten seconds"—and we were lying there, very tense, in the early dawn, and there were just a few streaks of gold in the east; you could see your neighbor very dimly. Those ten seconds were the longest ten seconds that I ever experienced. Suddenly, there was an enormous flash of light, the brightest light I have ever seen or that I think anyone else has ever seen. It blasted; it pounced; it bored its way into you. It was a vision which was seen with more than the eye. It was seen to last forever. You would wish it would stop; altogether it lasted about two seconds. Finally it was over, diminishing, and we looked toward the place where the bomb had been; there was an enormous ball of fire which grew and grew and it rolled as it grew; it went up into the air, in yellow flashes and into scarlet and green. It looked menacing. It seemed to come toward one.
>
> A new thing had just been born; a new control; a new understanding of man, which man had acquired over nature.
>
> That was the scientific opening of the atomic age.

At that time, Oppenheimer recalled some lines from the *Bhagavad Gita:*

> If the radiance of a thousand suns
> Were to burst at once in the sky,
> That would be like the splendor of
> the Mighty One. . . .

Rabi: The Modern Age

I am become Death,
The destroyer of worlds.

In the fall of 1944, while Rabi was still at MIT, rumors began to circulate that he might win that year's Nobel Prize. "Some people had told me that there was a possibility I might get it," he said. "I didn't take it very seriously. Then one afternoon in early November, I had a call, and this man said, 'I am Mr. Johnson. I represent a Swedish newspaper.' I said, 'Yes, Mr. Johnson?' He said, 'I suppose you can guess what I'm calling about?' I said, 'Yes, Mr. Johnson.' He said, 'Have you heard anything?' I said, 'No, Mr. Johnson.' He said, 'Neither have I.' Well, this was an introduction to a good night's sleep. The next day, the telegram arrived from New York, and that night it was announced over the radio that I had won the Nobel Prize for my work on the magnetic properties of nuclei. They hadn't given out any prizes since 1939, and so now they announced that the 1943 prize would go to Stern [Otto Stern, Rabi's mentor, honored for his measurement of the magnetic moment of the proton] and the 1944 prize to me. I was really very pleased. It was an enormous pleasure, and an excuse for many parties in different parts of the country—the Radiation Lab and Los Alamos and so on. I had a great time, and I can say, with only slight exaggeration, that I didn't draw a sober breath for weeks. I couldn't go over to Sweden—after all, the war was still on—but there was a sort of ersatz ceremony in New York, at the Waldorf-Astoria. There was a big dinner, and there were a few guests from the Scandinavian nobility. My medal was presented to me at Columbia by Nicholas Murray Butler. Poor fellow, he was eighty-two years old and was really pretty far gone—he was then blind. But he had a good, resonant voice, and made a few appropriate comments, during which he referred to me as 'Fermi.' " Enrico Fermi had come to Columbia from Italy in January of 1939, having won the Nobel Prize for physics the year before. "It was very interesting at Columbia. I had all my promotions quite rapidly—from student to instructor, and the rest—and although, by that time, I had been on the faculty for fifteen years, I had never once met President Butler or exchanged a single word with him. There were many people who were in and out of the president's house all the time—very social—and never got anywhere. It shows that at Columbia, at least for the period I was there, things were very fair."

Rabi was forty-six when he won the Nobel Prize, and I was curious to know whether he felt that winning it at a relatively early age could be harmful to a scientific career. He answered, "Yes. There are many examples. I think it can be a very useful thing to have, but it subjects the individual to enormous pressures. The prestige is fantastic—although, after all, it represents only a selection of the Royal Swedish Academy of Sciences. There are other academies as great. But the Nobel Prize does have this strong public appeal. There is some romance attached to it. It's like winning a huge lottery, except that you don't exactly compete—apart from the fact that once you enter one of the professions involved, you are, in a certain sense, competing all the time. It puts the winner on a sort of pedestal, because of the great public attention and prestige and also the prestige among one's colleagues. So that unless you are very competitive you aren't likely to function with the same vigor afterward. You know, it's like the lady from Boston who said, 'Why should I travel when I'm already there?' The prize also attracts you away from your field, because other avenues open up. I was never very competitive about the Nobel Prize, but when it did come it certainly offered many opportunities, and rather attractive opportunities, not to do physics, which is, after all, terribly difficult. I would not say that it helped me in my scientific development. On the other hand, it enhanced my activities of a more public nature. It will, in the long run, be up to the man who writes my obituary to say if these efforts were any good. But the Nobel Prize certainly enhanced them. Some people, like Enrico Fermi, it hardly affected at all. Others, like Harold Urey, became public figures for a time. Some of the science that I might have done later, I have to admit, was done by others, perhaps not with the same personal quality I always tried to inject in all my scientific work. But, anyway, it was done. So I can't say that the progress of science was impeded at all by my getting the Nobel Prize. I have always advised young scientists to stay away from government work, advisory committees, and the like—to attend to their discipline until they no longer possess the vast physical energy that is needed to do, say, good physics."

This brought up a subject that has always fascinated and puzzled me; namely, the aging process in scientists, and especially in physicists and mathematicians. In these disciplines, people tend to lead an oddly telescoped professional life, with the great achievements being made near the beginning. It takes extraordinary personal wisdom to adjust to this and to accept the fact that at a relatively early age one's primary career may be either finished or considerably diminished. In his later years, Einstein remarked to Oppenheimer, "When it has once been given you to do something rather reasonable, forever afterward your work and life are a little strange." I asked Rabi at what age, in his opinion, physicists tend to run down. "It very much depends on the individual," he answered. "I've seen people run down at thirty, at forty, at fifty. They may still do good work, but not like what they did before."

"What is the explanation, exactly?" I asked.

"I think it must be basically neurological or physiological," Rabi said. "The mind ceases to operate with the same richness and associations. The information-retrieval part sort of goes, along with the interconnections. I know that when I was in my late teens and early twenties the world was just a Roman candle—rockets all the time. The world was aglow. You lose that sort of thing as time goes on. It's the sort of thing that you want to hang on to if you can. And physics is such an out-of-the-world thing. It's not like history or poetry, or even painting. In them you never really lose contact with the world—it's right there before you. But physics is an otherworldly thing. It requires a taste for things unseen, even unheard of—a high degree of abstraction and a sort of profound innate philosophy. These faculties die off somehow when you grow up. You see them in children, who are fantastically interested in making things and in asking, 'Why? Why? Why?' Then, at a certain age, the children just become adults and are no longer very deeply interested in anything, except in the process of making a living and in sex and power. Money. Otherwise, they're not terribly interested. Profound curiosity happens when they are young. I think physicists are the Peter Pans of the human race. They never grow up, and they keep their curiosity. Once you are sophisticated, you know too much—far too much. Pauli once said to me, *'Ich weiss viel. Ich weiss zu viel. Ich bin ein Quantengreis'*—'I know a great deal. I know too much. I am a quantum ancient.' "

Alamogordo suddenly transformed physics from a private academic discipline to one with enormous political and moral implications. The few physicists who were there, like Rabi, understood that fact almost immediately after the explosion. "At first, I was thrilled," Rabi told me. "It was a vision. Then, a few minutes afterward, I had gooseflesh all over me when I realized what this meant for the future of humanity. Up until then, humanity was, after all, a limited factor in the evolution and process of nature. The vast oceans, lakes, and rivers, the atmosphere, were not very much affected by the existence of mankind. The new powers represented a threat not only to mankind but to all forms of life: the seas and the air. One could foresee that nothing was immune from the tremendous powers of these new forces. My own concern was to join in any efforts to contain these dangers."

Almost no one had known about the bomb project while the bomb was being developed. Very few people in the military had known, and even Harry Truman, who was then vice-president, had not known. General Groves wanted to keep the test a secret, even though the explosive flash would be seen for 150 miles. An elaborate cover story was prepared. It was a chemical explosion, an accident, some experiment that had gone wrong. The military people had even prepared a casualty list. (The actual release never made use of this list, but announced that an ammunitions depot had blown up and that there were no casualties.) One of Groves's assistants, who was told to insure the security of the test, remarked that it was like keeping the Mississippi River a secret. Of course, after the bomb was exploded over Hiroshima there were no secrets.

To physicists of my generation, coming into physics just after the war, the use of the bomb on two Japanese cities seemed unnecessary. It appeared to us that the Japanese had already lost the war. The moral issues looked to us, in hindsight, so clear. But we were not involved, and couldn't really put ourselves in the place of the people who were. I asked Rabi if he could remember how things had appeared to him at that time. He said that before the Germans surrendered, American physicists who were working on the bomb were desperately concerned that the Germans might make the bomb before the United States did. They knew exactly who the German nuclear physicists were, but they had no way of knowing how far the Germans had gone. Then, late in the war, they found out that the Germans had never really gone anywhere. Once American scientists had produced the bomb, it was more or less taken out of their hands. No scientist was asked to advise the military on what to do with the bomb. On one occasion, after the war, Oppenheimer, accompanied by Under-Secretary of State Dean Acheson, went to see President Truman. He expressed some regrets to Truman about having made the bomb. "Mr. President, I have blood on my hands," he said. Later, Truman said to Acheson, "Don't you bring that fellow around again. After all, all he did was make the bomb. I'm the guy who fired it off." However, within the Los Alamos community, before Hiroshima, there was an enormous amount of anguished discussion, especially about the possibility of using the atomic bomb in some sort of non-lethal demonstration in the presence of the Japanese.

Rabi did not see at that time how the bomb could be used in a demonstration. "Who would they send and what would he report? You would have to tell him what instruments to bring, and where to stand, and what to measure. Otherwise, it would look like a lot of pyrotechnics. It would take someone who understood the theory to realize what he was seeing. It was not a trivial point. You would have to have built a model town to make a realistic demonstration. It would require a level of communication between us and the Japanese which was inconceivable in wartime, and while they argued for weeks, or maybe months, over the meaning of the explosion, we would be honor-bound to wait for an answer. And what would President Truman say to the American people afterward? How could he explain to them that he had had a weapon to stop the war but had been afraid to use it, because it employed principles of physics that hadn't been used in warfare before?" In any event, the bomb was used, and four weeks after the Alamogordo test the war was over.

Rabi returned to Columbia and became chairman of what was a badly depleted Physics Department. "We had lost Fermi and many others," he told me. "They all went to Chicago. Robert Hutchins—then chancellor of the University of Chicago—did a very adroit recruitment job. He promised them the world. We had gone from being one of the best departments in the world to almost nothing in terms of senior people. We had very promising younger people, like Lamb." Willis Lamb shared the Nobel Prize for physics in 1955 with Polykarp Kusch, who had left Columbia during the war and then returned. "I set out to bring in some senior people. Nobody came. I was offering people jobs at $5,000 a year more than my own salary, and I had the Nobel Prize—and yet I was being turned down. I never understood it. I was teaching a lot of the theoretical courses myself, and would bring people in on a part-time basis to teach some of the specialized courses. Kramers, for example, was sent to the United Nations by Holland for a debate on atomic energy, and he came to Columbia at night to teach quantum mechanics." H. A. Kramers was one of the most distinguished of the Dutch quantum mechanicians. "Finally, I decided that we would have to begin by developing our own young people and bringing in other young people. I felt that what I could do was to make sure they would have facilities that otherwise they might not have. New fields were opening up, like the study of elementary particles—the mesons, and so on. So I got money from the Navy to build what was then a large accelerator. I knew that if we had such a facility we would attract young people who wanted to work in the new areas. My own group in molecular beams—an extraordinary collection of people—was completely disrupted by the war. I never did succeed in building it up again to where it had been. One of the reasons was that I wanted the department to broaden, so I did not try to hog the best students. In fact, I got very few of them. My own research efforts were a small proportion of the whole thing. But we became a very great physics department once again." Since the war, five present or former members of the Columbia Physics Department have won the Nobel Prize for work done during that period: Lamb and Kusch in 1955; T. D. Lee in 1957, jointly with C. N. Yang, for their work on the non-conservation of parity; Charles H. Townes in 1964, jointly with the Russian physicists Nikolai G. Basov and Aleksandr M. Prokhorov, for their work leading to the development of the maser and the laser; and the latest being J. Rainwater in 1975 for his work in theoretical nuclear physics.

At the end of the war, no one knew quite what to do with the atomic bomb and nuclear energy. There were two opposed points of view. Within the government and the military there was a strong feeling that the essential thing was to maintain the American advantage in nuclear energy by keeping American work completely secret from the rest of the world. But within the scientific community there was an equally strong feeling not only that this was impossible but that we had a wonderful opportunity within the context of the newly formed United Nations by pooling a large part of our nuclear resources under international control. The administration's view was embodied in a bill introduced in Congress by Senator Edwin C. Johnson, of Colorado, and Representative Andrew J. May, of Kentucky. The bill looked superficially satisfactory to the scientists, but as they came to understand more and more of its implications it began to appear worse and worse. It stipulated that a commission was to be appointed by the president to assume complete control of nuclear energy. There was no provision for a balance of civilian and military interests, and scientists feared that members of the military might dominate the commission. The whole tenor of the bill was in the direction of secrecy and, so it appeared, the military control of nuclear energy.

Until that time, the physicists and the others who had worked on the development of the bomb had never really tried to act communally as a political unit. But the May-Johnson bill brought them together. Led by such people as Leo Szilard, who had been one of the early researchers in nuclear fission, and E. U. Condon, who had been at Los Alamos and was now associate director of the Westinghouse Laboratories, they began a lobbying and educational campaign against the bill. The Federation of Atomic Scientists was formed, and speakers were sent all over the country to clubs, to church groups—to anyone who would listen—to try to educate people about the dangers of an autocratic commission whose members might end up being drawn entirely from the military. By December of 1945, enough opposition had been generated within the Senate to kill the May-Johnson bill. It was replaced by the McMahon bill, sponsored by Senator Brien McMahon, of Connecticut, and that bill passed in the summer of 1946. It created the AEC—a civilian body. On October 28, 1946, President Truman appointed the first AEC, of five members, with David Lilienthal as chairman. Robert F. Bacher, a former Los Alamos physicist, was the scientific representative. None of this put the nuclear genie back into the bottle. The bomb was, and is, still with us. But it did mean that the scientific community had succeeded in shifting the emphasis more toward private research and the peacetime application of nuclear inventions.

The control of the atomic-energy program passed from General Groves to the civilian AEC on January 1, 1947. Soon after that, the General Advisory Committee of the AEC began its sessions. This was a powerful group, also selected by the president, whose purpose was to advise the commission on scientific and technical matters. Oppenheimer was elected chairman—it was all but inconceivable to the people involved at that time that anyone else would be chairman—and Rabi was soon asked to join it, and he did. Rabi told me, "I was very, very glad to do it, because I thought that by working from within, so to speak, we might be able to do something about getting rid of the atomic bomb. So I never participated in the external demonstrations—the newspaper ads, and things of that sort. It was my feeling that you had to be inside the government if you wanted to have an influence, especially on these military matters. Since there was all that secrecy, you couldn't know what you were talking about unless you were a part of it. So much in that business depends on the specific nature of the weapons—the processes, the means of delivery, some idea of the future developments and of what the other side is doing. I was all in favor of the people outside the government clamoring in a kind of blunderbuss way against the spreading of atomic weapons, and secrecy, and so on. But I thought that I could be more effective by working within."

Even before the AEC was created, the atomic scientists had been considering the prospects for international control of atomic weapons. Rabi said, "Oppenheimer and I met frequently and discussed these questions thoroughly. I remember one meeting with him, on Christmas Day of 1945, in my apartment. From the window of my study we could watch blocks of ice floating past in the Hudson. We were then developing the ideas that became the basis of the Acheson-Lilienthal Report"—it was issued in March of 1946—"and, later on, the Bernard Baruch position, which the United States took in the United Nations. It was a most extraordinary kind of liberal position for the United States to take—that there be no private ownership and no national ownership of uranium and other fissionable materials. They should all be internationally owned under United Nations supervision. This was the actual American position, at a time when we had a monopoly on the atomic bomb."

In reading the Acheson-Lilienthal Report now—especially in the light of the nuclear-arms race, which appears to be spreading everywhere—one is struck by its somber prescience. It has all the literary characteristics of Oppenheimer's style, and most people believe that it was largely written by him. A section entitled "The Elimination of International Rivalry" begins:

> It is clear that uranium and thorium are materials of great strategic importance to nations seeking to establish for themselves a powerful position in the field of atomic energy. The fact that rich sources of such materials occur in a relatively few places in the world, as compared, for example, with oil, creates a competitive situation which might easily produce intolerable tensions in international relations. We believe that so long as nations or their subjects engage in competition in the fields of atomic energy the hazards of atomic warfare are very great indeed. We assume the General Assembly of the United Nations, in setting up an Atomic Energy Commission, had this disturbing fact much in mind.
>
> What is true in respect to the dangers from national competition for uranium is similarly true concerning other phases of the development of atomic energy. Take the case of a controlled reactor, a power pile, producing plutonium. Assume an international agreement barring use of the plutonium in a bomb, but permitting use of the pile for heat or power. No system of inspection, we have concluded, could afford any reasonable security against the diversion of such materials to the purposes of war. If nations may engage in this dangerous field, and only national good faith and international policing stand in the way, *the very existence of the prohibition* against the use of such piles to produce fissionable material suitable for bombs would tend to stimulate and encourage surreptitious evasions. This danger in the situation is attributable to the fact that this potentially hazardous activity is carried on by nations or their citizens.
>
> It has become clear to us that if the element of rivalry between nations were removed by assignment of the intrinsically dangerous

phases of the development of atomic energy to an international organization responsible to all peoples, a reliable prospect would be afforded for a system of security. For it is the element of rivalry and the impossibility of policing the resulting competition through inspection alone that make inspection unworkable as a sole means of control. With that factor of international rivalry removed, the problem becomes both hopeful and manageable.

The American proposal for international nuclear cooperation was submitted to the United Nations Atomic Energy Commission by Baruch on June 14, 1946, but almost immediately Andrei Gromyko, on behalf of the Soviet government, began formulating objections to the plan, and it was eventually defeated. The reasons for the Soviet stand were certainly complex. It is quite likely that the Russians were, even at that time, well along with their own program for developing nuclear explosives—something that they may have begun during the war. (It is not generally realized that the British, too, had begun working on the development of nuclear fission before the organization of Los Alamos—indeed, long before the bomb project was taken seriously here. They were unable to proceed with the development, because they could not spare the resources during the war.) Besides, the Russians distrusted the Americans' motives. Rabi recalled conversations that he had at that time with Dmitri Skobeltsyn, a Russian physicist, who was with the Soviet United Nations delegation, and who came to see Rabi at Columbia informally. "Skobeltsyn was a good physicist and an intelligent man," he told me. "He would come to me with a copy of some fire-eating speech made by some senator. I would try to tell him not to pay any attention to it. But how could he believe me? I think they just didn't trust the whole thing. They thought that it was some capitalist trick and that we wouldn't adhere to the agreements. They felt that it was all a way of stopping them, and this belief was probably reinforced by the rather belligerent way in which Baruch actually presented the proposal. Or I think they felt that, and certainly it was natural for them to feel that." In any event, the nuclear-weapons race was on. In the summer of 1949, the Russians exploded their first atomic bomb, and the world has been living with the proliferation of nuclear weapons every since.

On September 23, 1949, President Truman announced the Soviet atomic explosion. (After he left office, he indicated that he hadn't really believed that the Russians had exploded an actual bomb.) Among the people concerned with our own program, the Soviet test precipitated a crisis. Several influential physicists, including Edward Teller, of Los Alamos, and Ernest Lawrence and Luis Alvarez, of Berkeley (Lawrence won the Nobel Prize in physics in 1939, for having invented the cyclotron; Professor Alvarez won the prize in 1968, for his development of techniques for the detection of subatomic particles), felt that some decisive countermeasure was necessary, and that our government should begin a crash program to make "the Super"—a hydrogen bomb—since making such a bomb would guarantee our continuing vast superiority. It was the complex of events surrounding President Truman's decision, announced on January 31, 1950, that the AEC "continue its work on all forms of atomic weapons, including the so-called hydrogen or superbomb," which led, four years later, to Oppenheimer's security-clearance hearing.

A great deal has been written about how the decision to make the Super was reached. Rabi, as a member of the General Advisory Committee, participated in the decision. I asked him what his recollections were of what had happened, and he said, "The Atomic Energy Commission was asked to give the president advice on whether or not to initiate a crash program to make a hydrogen bomb, and it asked the Advisory Committee." At that time—October of 1949—the members of the Advisory Committee, in addition to Rabi and Oppenheimer, were James Conant; Lee DuBridge; Enrico Fermi; Hartley Rowe, an engineer who had worked on materials procurement for the Manhattan Project; Cyril Smith, a metallurgist who had been in charge of the metallurgy division at Los Alamos during the war; Oliver E. Buckley, president of Bell Laboratories and an expert on guided missiles; and Glenn Seaborg, of Berkeley, a noted chemist who shared the Nobel Prize for chemistry with E. M. McMillan in 1951, for their discoveries of the transuranium elements. "It was a real crisis time, and the whole problem was so entangled that it is difficult for me to remember all the ins and outs of our debate. In fact, the debate was recorded but the tapes were deliberately destroyed soon afterward. It is a great pity that we cannot hear the voices of Fermi and Oppenheimer and the rest in that very fateful discussion," Rabi went on. "The problem that came up was this: All through the period at Los Alamos, and before, there had been a proposal for a bomb called the Super, on which work was being done. The general idea was to cause a fusion reaction involving hydrogen." Hydrogen nuclei (protons) are "fused" by confining them in close proximity at exceedingly high temperatures. When the fusion takes place, the resultant nucleus—deuterium, for example—is less massive than the sum of the masses of its constituents. This mass loss is converted into energy according to Einstein's formula $E=mc^2$, m in this case standing for the mass loss. "It was an interesting idea, except that whatever specific theoretical proposal was made didn't seem to produce a self-propagating chain reaction," Rabi went on. Such a reaction was necessary to generate enough energy to cause a large-scale explosion. "Beautiful and clever calculations were made about it, and even more brilliant calculations to show that it wouldn't work. It began to be more and more evident that if there were such a device it would have to be enormous. A ship would be needed to hold it. And then there was the question of how it would be set off. Would the fusion process spread—'propagate'—throughout the nuclear fuel or simply confine itself to a small part of it? Would it cool off after you got it ignited? Since it would be at a very, very high temperature, would the device radiate away that heat and cool off before it could explode? Nevertheless, after the Russian explosion, certain people thought that we did need some counterthrust. And Lawrence, Alvarez, and Teller felt that the only thing to do was to go full tilt for this Super. In fact, during the Los Alamos period some people felt that it should have higher priority than the ordinary fission bomb. 'We shouldn't dillydally with the fission bomb but go for the Super'—so said some very eminent physicists. Not very sensible, but very eminent. I wonder what they think about that now. Anyway, after the Russian explosion they wanted to go ahead with a specific Super model. This model is what came up before the committee. There were two objections to it. In the first place, it was a very chancy thing, because, basically, we didn't know how to make it. And then, just about this time, it was shown that it wouldn't work, it wouldn't propagate—at least, not that particular model. But the general kind of thing they were talking about would have been absolutely devastating if it had worked, because it was so big. Well, the Super people pressed the Atomic Energy Commission to give this program first priority. They wanted to try various configurations of this general type. There was no way we had of refuting these models in general. A man could come in with some configuration and be strongly in favor of it, and it might take three months or more of calculations, by some very brilliant people, to find out if it had a chance. There were no experiments, and the constants weren't known, or anything of that sort. It was like dealing with some of the perpetual-motion people. You show them why something doesn't work, and they say, 'Fine. Thank you,' and the next day they're back with a modification.

"In any event, there was strong agreement within the committee that we should not go ahead. We all agreed that if it could be made to work it would be a terrible thing. It would be awful for humanity altogether. It might give this country a temporary advantage, but then the others would catch up—and it would just louse up life. Fermi and I said that we should use this as an excuse to call a world conference for the nations to agree, for the time being, not to do further research on this."

Rabi and Fermi wrote, in an addendum to the full committee's report to the AEC, "The fact that no limits exist to the destructiveness of this weapon makes its very existence and the knowledge of its construction a danger to humanity as a whole. It is necessarily an evil thing considered in any light. For these reasons we believe it important for the president of the United States to tell the American public and the world that we think it is wrong on fundamental ethical principles to initiate the development of such a weapon."

"Fermi and I felt that if the conference should be a failure and we couldn't get agreement to stop this research and had to go ahead, we could then do so in good conscience," Rabi said. "Some of the others, notably Conant, felt that no matter what happened it shouldn't be made. It would just louse up the world. So the committee unanimously agreed not to go ahead with it with this high priority, especially since we were doing very well in the development of the fission bomb. One member was absent—Glenn Seaborg—and I don't know how he would have voted if he had been there for the discussions. Well, the committee's report to the AEC caused a lot of flak—a lot of consternation—in some circles, and especially with the Berkeley group and Teller. I think we persuaded some members of the commission, but one, Lewis Strauss, got into an absolute dither. The Berkeley people and Strauss went around and talked to newspapermen all over, and to the House and the Senate, and whatnot. Furthermore, it was just about then that Klaus Fuchs was arrested in London for espionage on behalf of the Soviet Union." Fuchs had been at Los Alamos from December of 1944 to June of 1946, working on, among other things, the primitive Super program. He was in a unique position to have complete knowledge of American efforts up to that point to make a Super. "All of this got to President Truman and built up such a head of steam that he was practically forced to declare that he was going to give the Super top priority. Now, around this very time, some experiments that were suggested by Teller and Ulam"—this was Stanislaw Ulam, a brilliant Polish-born American mathematician, who was at Los Alamos during and after the war—"gave rise to a new idea for making a thermonuclear reaction. It bore no relation to the original Super, and in a fairly short time it was shown to be a practical thing, which could be calculated. And the General Advisory Committee backed it—with reluctance, because of all the problems it would create. But it was not the horrendous first thing. It was a terrible thing, but not the original Super. It came not long after Truman's original decision, so the two devices got mixed up in people's minds. Now, I never forgave Truman for buckling under pressure. He simply did not understand what it was about. As a matter of fact, after he stopped being president he still didn't believe that the Russians had had a bomb in 1949. He said so. So for him to have alerted the world that we were going to make a hydrogen bomb at a time when we didn't even know how to make one was one of the worst things he could have done. It shows the dangers in this kind of thing. He didn't have his own scientific people to consult and give him an impartial picture."

In 1952, Oppenheimer's term of office as chairman of the General Advisory Committee expired, and Rabi became chairman. In addition, Rabi played a key role in developing the concept of a special science advisory service that would be responsible to the president and work directly with him. The concept had had its origins the year before in the Truman administration. A Science Advisory Committee was set up under the direction, at first, of Oliver Buckley, from the General Advisory Committee of the AEC. The Science Advisory Committee met in the Office of Defense Mobilization once every few months. Rabi joined the group after it had been in existence for about a year. Rabi remembers going to see Truman and telling him he was about to write an article that would contain some ideas about lifting secrecy in the atomic-weapons field which might conflict with the president's own ideas. Recalling this visit, Rabi told me, "Truman said, 'I don't muzzle anybody. You go ahead and write what you think.' I didn't have much actual talk with him, but I was impressed by how direct a man he was. I could have seen much more of him if I had wanted to, but all the time I was in government I believed in working through channels. Since I was a professor at Columbia and had my job here in New York, I felt that if I tried to do something by going over the people's heads I wouldn't get anywhere. The full-time people were in Washington, and here I was in New York. I might have had some success in one thing or another, but then the man on the job can do all sorts of lobbying while you are away. Going over people's heads didn't seem like a sensible thing to do unless I wanted to move to Washington and be a full-time adviser, which I never had any desire to do."

To the surprise of many physicists, Rabi publicly supported Dwight Eisenhower in both his election campaigns against Adlai Stevenson, whose intellectual qualities appealed to most scientists. I asked Rabi if subsequent events had caused him to change his views. "No," he said. "I think Eisenhower, when it's really added up, will turn out to have been a great president in those difficult times. All I can say is that in the eight years of his presidency, in spite of fantastically wonderful opportunities to go to war, and in spite of having Mr. Dulles around, who was a real fire-eater, and having Mr. Nixon as vice-president, he managed to evade it. As soon as he stepped out, President Kennedy couldn't resist these confrontations—the beginnings of the Vietnam War—and, of course, President Truman, before him, couldn't resist going to Korea."

Rabi had an especially good rapport with Eisenhower, whom he had known from the years between 1948 and 1952, when Eisenhower was president of Columbia. "It was possible to talk openly with Eisenhower," he said. Rabi helped to convince President Eisenhower to create the position of special assistant to the president for science and technology. "The job of the special assistant was to understand the president, his personality, his strengths and limitations, and then to try to advise him and anticipate his problems, if possible, and to keep the president interested," Rabi said. "The job was not to be a great man around the place but to understand the president's problems and be helpful. During this period, I was chairman of the president's Science Advisory Committee, so I was able to talk to Eisenhower about this. I told him that he should pick a person he liked and could get along with, who could be his confidant in many matters. The president himself cannot be expected to see the scientific aspects of the problems that come before him. He needs someone who is with him on an almost day-to-day basis. Otherwise, the various agencies of government—the Department of Defense and the other departments—can tell him anything, and he has no recourse of his own. If he had his own science adviser, he could say, 'Look at what these people are telling me. Can you find out more about it?' Then, through the science adviser and the president's Science Advisory Committee, the best scientific minds in the country could be marshaled for such problems on a purely ad-hoc basis and report to him. Furthermore, if he had some worries about how the departments were doing—particularly the Department of Defense, spending all that money and with all that technology—he could ask the adviser to look into it. I felt that all this was very important, and later Eisenhower said that he didn't know how he lived without it before—otherwise, he really couldn't know about these technological things. No president can. That's what so astonished me about President Nixon. He had this instrument working for him, and he just threw it away. It was unbelievable. To this day, I can't understand the thought processes that went into that—except that it may have been that clique around him, who were very intolerant about dissent from their views."

The first special assistant for science and technology was appointed in 1957, at the time of Sputnik. He was James Killian, the president of MIT. He resigned in 1959 and was succeeded by George Kistiakowsky, a chemist from Harvard, who, under President Kennedy, was succeeded by Jerome Wiesner. Rabi told me that as time went on he felt that the original purpose of the post, and of the committee, had been lost sight of. "The president's Science Advisory Committee went into more subjects than I thought we could really handle," he said. "Some of the members felt, for example, that we should do something about science education, and we set up a panel for education, and so on. We set up an Office of Science and Technology, with not a big staff—but still a considerable staff. And after we got what we thought were competent people into the scientific side of the Department of Defense, we took less interest in that side of things. So in a sense we moved away from and lost the ideal that we had started with—not to try to make policy but to advise the president. This narrow thing of advising the president. But it was effective anyway. People inside and outside the government who needed support for their scientific work had a representative—people, for example, from the Department of State or the Department of Commerce. So the committee may have moved away from its special function of advising the president, but it did represent a strong central focus for science. It raised the level of science in all the departments. Most of the cabinet members came to have some sort of assistant secretary for science and technology. And another result was that most countries with whom we had to deal—England, France, Germany, and so on—set up ministries of science and technology along this general line. Of course, there was no way of making a president like his science adviser once he had been appointed, but still he didn't want to fire him. So he made less and less use of him and of the Science Advisory Committee, too. This is what happened during the Johnson administration, and we had very little contact with Johnson. The committee was still useful, but not on this narrow thing of giving advice to the president. We were never once consulted, for example, on the war in Vietnam."

The beginning of the Eisenhower years was a bleak time for most American physicists. The Oppenheimer security-clearance hearing came at the height of the McCarthy period, when it was not at all clear that civil liberties—not to mention science—were going to survive in this country. The hearing itself began at the headquarters of the AEC on April 12, 1954. It was conducted by a special panel of the commission, and the ostensible issue that the panel was to decide was whether or not to declare Oppenheimer a "security risk" and hence to deprive him of his clearance to work on classified material. In reality, essentially all the supposedly negative material brought out against Oppenheimer had been reviewed in numerous prior security hearings, in which Oppenheimer had been cleared. This fact, combined with the extralegal quality of the proceedings (courtroom rules of evidence were rarely followed), outraged most scientists—as did the absurd idea that Oppenheimer could be "walled off" from a set of ideas that he as much as anyone was responsible for creating. Nonetheless, on May 27, 1954, the panel decided against Oppenheimer, thus ending his government service and casting a shadow over him—a shadow from which he was never able to emerge. In 1957, when I was a visiting member of the Institute for Advanced Study, of which Oppenheimer was then the director, I once, by chance, took a train ride with him from Princeton to New York. I was not much more than a graduate student and, of course, I hardly knew Oppenheimer, but for some reason he began telling me about what he referred to as his "case." The principal thing that stayed with me from that conversation was his remark that throughout the hearing it felt to him as if it were happening to someone else. In a sense, he found it so unbelievable that, certainly at the beginning, he did not take it seriously—perhaps not seriously enough. But even if he had taken it seriously the outcome, in all probability, would have been the same.

Rabi was one of the strongest witnesses in Oppenheimer's defense. That testifying for Oppenheimer might prejudice his own governmental role was of no concern to him. At one point in the hearing, he was told—and this was fairly typical of the level on which the hearing was conducted—that "perhaps the board may be in possession of information which is not now available to you," but with no indication of whether or not there was such information or in what area it might be. Rabi agreed that there might be such information—some incident not specifically known to him—and then he said, "On the other hand, *I* am in possession of a long experience with this man, going back to 1929, which is twenty-five years." Any such specific piece of information, he pointed out, must be taken together with the "sum." And he went on, "That is what novels are about. There is a dramatic moment in the history of the man—what made him act, what he did, and what sort of person he was. That is what you are really doing here. You are writing a man's life." Unfortunately, that was not what the board was really doing. It was trying to get Oppenheimer out of government service with a maximum of disgrace, so that his influence would forever after be diminished—at least in the government. In this, the board succeeded completely.

It is an element of high tragedy that the selection of the victim is, somehow, never completely accidental. There were aspects of Oppenheimer's character that played into the hands of this kind of inquest. I asked Rabi for his opinion of how Oppenheimer had become vulnerable. Rabi said, "When I first met him, in late 1928, he was absolutely apolitical. He was an aesthetic man. He even contributed to *The Hound & Horn*—a very highbrow magazine that was then being put out by some group at Harvard. But in the Hitler period he sort of became interested in politics. I think he had some relatives in Germany. So, somehow or other, perhaps through friends, he got interested in teachers' unions and things like that. And you know Oppenheimer—once he got interested in something, he went right in to become the leader of it. Just as it was in physics—he would sit there in Princeton like a spider in the center of a web, with all sorts of lines of communication everywhere. I remember one time I was visiting Berkeley and I went up to a group of physicists and said as a joke, 'I see you have your genius costumes on.' The next day, in Princeton, Oppenheimer knew that I had said that. You can see that once he got into something he'd be in it with both feet. He was practically running the teachers' union at Berkeley in the thirties, with its left-wing connections. He became quite an expert—I don't know how much of one—on Marxist theory. In those days, he was sort of philosophical, deeply interested in religious things bordering on mysticism. I think he never lost that interest. He had that streak, which could sometimes be very foolish."

I remarked to Rabi that some of the things that emerged during Oppenheimer's hearing seemed to me to be the product of unfortunate human judgments on Oppenheimer's part.

"Sometimes judgments, and sometimes he just liked to tell tall stories," Rabi answered. "His being called for a hearing at all was due partly to his tall stories. He hated not to be in on anything. He was a very adaptable fellow. When he was riding high and had support, there was nobody more arrogant. When he felt that things were against him, he could play the victim, and he did. If you read the transcript of the hearing, you feel he was the victim. If you read some other things, like his testimony in the Senate, you find he was fantastic. You could see that the man was in complete control. He would answer a question in such a way as to provoke the next question, for which he was ready. It's a wonderful thing to read—he was a most remarkable fellow. After the hearing, I was asked into government service more than I really wanted to be. I could have got into a position of very great influence, which I didn't want. I was kind of modest about the amount of wisdom one could contribute on a part-time basis, and I was not ready to take one of those responsible administrative jobs. When Oppenheimer retired from the chairmanship of the General Advisory Committee of the Atomic Energy Commission, in 1952, I became chairman, and I remained chairman for four years and then retired. That position gave me the opportunity to contribute to what I think was the most important thing that we did, and was the real basis of our détente with Russia—organizing the International Conference on the Peaceful Uses of Atomic Energy."

Rabi told me that Dag Hammarskjöld had been of the opinion that the Conference on the Peaceful Uses of Atomic Energy, which took place in Geneva in 1955, was the most important diplomatic event of the 1950s. "It made a very great difference in the whole idea of atomic secrets," Rabi said. "We and the Russians were forced to declassify a whole field of nuclear physics and technology in order to take a position at the conference. Declassifying the papers one by one would have taken forever. We simply opened up the field. There was an exchange of technical information that came to more than twenty volumes. It was the first time the Western scientists and the Soviet scientists had met together in an important international cause. It introduced a human as well as a scientific connection. Unlike the usual cultural exchanges—like sending a ballet troupe to Russia or getting the Bolshoi to come here, which I don't decry at all—this connected the positions of power to these very important issues. It was the first time that such a group had been even thinkable. The scientists met, they got to know one another, and in this way they were able to influence their own governments. I think it changed the whole direction of things. My job was to persuade Strauss, who was then chairman of the Atomic Energy Commission, to persuade Eisenhower to have such a meeting. We then had to get the State Department to put it forth as a proposal at the United Nations. We had to persuade other countries not to oppose it in the UN. My whole purpose was not just to get *people* together but to get the *governments* together—to get them to have a high stake in it, at a high level. It had to be a big conference. It had to have enormous publicity, and it had to involve national prestige. Unlike the papers at the usual scientific conferences, these papers were to be presented not as the work of individuals but by governments. I helped Dag Hammarskjöld select the seven countries that were to be on the organizing committee. We chose Canada, because it had a lot of uranium; Brazil, as a representative of the South American countries; India, which had some thorium sands, as the Asian representative; England and France; and, of course, Russia and ourselves. Then the committee met to decide what sort of conference it should be. The great danger was that it might lead to something terrible. So we had to have a conference that was run by countries, but with a kind of agreement that nothing political would enter the rules of procedure. That was a terrific thing to arrange. It took a full week for twenty-two simple rules of procedure to be agreed on. The Russians opposed us at every turn, and when the final vote came it was counted: 'One, two, three, four, five, six'—and then Hammarskjöld turned to the Russian delegation and said, 'You see, the vote is against you,' and the chief Soviet delegate said, 'But I haven't voted.' Hammarskjöld apologized profusely, and the Russians voted with the rest of the countries. To this day, I don't know whether the Russians' objections to the rules of procedure were an attempt to be able to insinuate politics into the conference or were a test of the sincerity of the West to hold such a conference. But, believe me, the effort was tremendous."

In 1964, Columbia appointed Rabi a university professor. He was the first to hold this special academic post, which allowed him complete freedom to do anything he wanted as far as teaching was concerned, and which he held for three years. Columbia professors must retire on the June 30th following their sixty-eighth birthday. Since Rabi was born on July 29, 1898, his retirement was on June 30, 1967, when he became a professor emeritus. "Retirement held no terrors for me," Rabi remarked not long ago. "I've been fortunate—had a very blessed kind of existence. I have seen what happens to older people—I can see that this kind of existence comes to an end. Although I have been basically retired, I have retained some connections with industry, many with government—things like the old president's Science Advisory Committee, the UN, NATO, and the International Atomic Energy Agency. I am a member of various boards of trustees—the Hebrew University, in Jerusalem, and The Mount Sinai Hospital, here in the city. I am not bored at all, and I still have this close connection with one of the great physics departments of the world. On my free days, I can go over there and walk down the hall, and some of the greatest minds in the world are right there. I have not devoted myself to science lately—it's just plain too hard. As you know, I keep somewhat in touch with it, but not in a creative way. I don't think I could do that. I'm always afraid of being a stuffed shirt—making do with pretense rather than actuality. I have been too close to really good science. There are some cases of people who have done good things after they got the Nobel Prize, but if you set that standard the chance of doing something like that again is not great, and it's very hard to be satisfied with less. You have to face it: part of it is accident. You were there with an idea at the right time, and so had the opportunity to do the work. Great scientists usually show up when and where in the history of science they are needed. Of course, people don't just appear out of a zero environment. They usually have had a teacher—somebody in that field, somebody in that tradition. But they can appear at the wrong time. I know perfectly well that there are people living now who could have created quantum mechanics, but they were born too late. It had already been done. Nobody's going to rediscover Newton's laws of motion—and, in that sense, neither would you expect Newton to do it again. Although you might say that he almost did. Einstein was beyond anybody; he did three great things—relativity, gravitation, and the quantum theory."

I asked Rabi if he had been doing much lecturing in the last few years.

"Not very much," he replied.

"Do you like to give lectures?"

"Once I am started, I don't mind," he said. "I used to like to do it. Now it's a matter of going to a place for a short time and meeting all sorts of people for a short time. There is a general futility about it. One is almost an entertainer. In some cases, people do remember a lecture, and something comes out of it. You work for those cases. But they are very few. No, I'm not the kind to lecture, although I have got a great deal out of lectures that I have heard other people give. I think lecturing is a fine thing, especially for Americans, because we learn through the ear. In most of our education, that is the method. An Englishman is more likely to read, but we are not a country of readers. We'll even tolerate lectures on subjects for which a textbook exists. In England or Germany, this is unheard of. If somebody has written it down, why go to a lecture? But Americans will go."

I asked Rabi what had particularly struck him in the development of physics since the Second World War, and he answered, "Two things. First, the wonderful developments that have come out of the old physics—the prewar physics—as a result of the new points of view. For example, the whole new development of maser and laser physics came from Einstein's theory of stimulated emission, which he presented in 1917." This is the theory that when atoms are in excited states electrons can be stimulated to emit quanta by the presence of other radiation quanta. "It was an idea then, and now this whole new field has come out of it. All the new things in solid-state physics—like the transistor—come from the older physics. Plasmas, fusion, and the whole development of reactors came directly from the prewar physics with a new, sophisticated understanding. But the really new field to me is the field of elementary-particle physics, in which a whole new world is being opened up—a more profound view of matter and anti-matter. Most of this elementary-particle physics was totally unknown before the war, and had to await the development of entirely new instruments, like the giant accelerators, which were constructed only after the war. While we have been operating here with old concepts but in new forms, I feel sure that there will be some entirely new insights and achievements. To me, that is the most exciting thing in the whole world of science. To me, that's No. 1. The biological discoveries, for example, like DNA, although they are of tremendous import—really marvelous—are understandable once they're stated. You can understand them on the basis of older ideas in physics and chemistry. But in the particle physics you have something entirely new—tremendously moving and mysterious.

"My view of physics is that you make discoveries but, in a

certain sense, you never really understand them. You learn how to manipulate them, but you never really understand them. 'Understanding' would mean relating them to something else—to something more profound. But these basic things in physics can't be related to anything more profound. Everything has to be related to *them*. So in that sense you can never really understand them. They become tools for a deeper understanding of the basis of experience and existence. The remarkable thing is that this brain, which I suppose developed so that we would not run into trees and would be able to keep ourselves alive somehow or other, is capable of this very high degree of abstract thinking and of the investigation into itself and its own meaning, into the meaning of life and what is beyond the visible. Still more, it has enabled us to create all these extensions of the human powers: telescopes, radio-telescopes, microscopes, with which one can probe way beyond the powers of ordinary human vision into the very constituents of matter, and computers, which have enabled us to reach conclusions that we couldn't have done with the human brain alone. A modern man has at his disposal fantastic, undreamed-of powers with which to investigate further.

"Physics now is not just a development of the physics of twenty years ago. It is quite new. It turns corners. There are always surprises. Not the sort of surprises you find in investigating a mechanism, as in physiology, but surprises that evoke ideas that did not exist before. I don't think that physics will ever have an end. I think that the novelty of nature is such that its variety will be infinite—not just in changing forms but in the profundity of insight and the newness of ideas that will be necessary to find some sort of clarity and order in it. I think the only thing you can compare it to is the mystical idea of God with infinite attributes. Physics is that sort of thing in a tangible form. In addition, you acquire modes of thought that apply to other things—not only to physics. It opens doors to new forms of thought. It has not had any real application in our society, because the people who might use it, in politics and so on, just have no idea about it. They have not been educated in what exists and what is at hand. It's a great pity that the general public has very little inkling of the tremendous excitement—intellectual and emotional excitement—that goes on in the advanced fields of physics."

"How can that gap be bridged?" I asked Rabi.

"I think only by drastic action, and I think this is necessary to preserve the human race," Rabi responded. "It may come—but not in my lifetime. I think that physics should be the central study in all schools. I don't mean physics as it is usually taught—very badly, as a bunch of tricks—but, rather, an appreciation of what it means, and a feeling for it. I don't want to turn everybody into a scientist, but everybody has to be enough of a scientist to see the world in the light of science—to be able to see the world as something that is tremendously important beyond himself, to be able to appreciate the human spirit that could discover these things, that could make these instruments to inquire and advance into its own nature. I rate this so highly that I have a feeling that with this education people would find something above their everyday lives, above their nationality, even above their religious affiliations, and find a basic unity in the spirit of man."

"Don't you think that the same sort of universality can be got by studying the humanities?" I asked.

"No," Rabi answered. "Nothing near it. The humanities cater to some of your senses. We have a tremendous feeling for words, for speech—we have to have it—and the humanities develop that. They cultivate emotions and set up a hierarchy of values, whatever those values may be. The values can be harmful, and a large fraction of them are harmful to the further existence of the human race—excessive patriotism and things of that sort. However much I enjoy the humanities—and I do, vastly; they enrich my life—I don't think they are truly universal. They are universal only in the sense that every group has some form of the humanities in its history, no matter how primitive the group. The humanities have the purpose of keeping small groups together and perpetuating some cultures. Also, we have moved more and more in the direction of humanism in the service of practicality. There is an old Jewish saying that one should not use the Torah as a spade. If you teach science or the humanities as a model of how you would win an argument or carry on a transaction, you lose the feeling—in science, at least—for the inner beauty. So much of science is taught as a prelude to practical use, but I think there is something sacred about it, and unless it is taught that way you will simply miss most people—all except those who have some innate urge toward it and are not going to be dissuaded by any amount of bad teaching."

I asked Rabi if he felt optimistic about the future—the future of the human race.

"Not very much, from what I see at present," he replied. "The human race will have more and more the capacity for blowing itself apart, and the miracle will be if that doesn't happen. If you look at the whole postwar period, as our powers of destruction have increased we have been skirting closer and closer to the edge, instead of working full-time to ease the situation. We keep returning to the old emotional bases of everything, as in the Cold War. We have defended to the death all sorts of secondary values, as in Vietnam—one regime over the other, when both were obviously bad. We have gone out of our way to defend certain principles that are not themselves well founded—not necessarily adapted to people. If you look through the whole postwar period, it has been one crisis after another, bringing us closer and closer to the ultimate danger—within arm's length. And I think the danger is going to spread. Not only from nuclear weapons but from all sorts of other devices. Nuclear weapons are not necessarily the ultimate in destruction. There is a whole realm of biology and chemistry that has great potential for doing harm.

"Also, I find the kind of leaders of these masses of humanity, whether here or in the Soviet Union or anywhere else, really distressing. Where are the great men who should be leaders of the world? Right now the leaders are nobodies. I find some kind of optimism in that I see people who will work very hard to give us time to stave off catastrophe with very little expectation of reward for themselves. It's on these people that the outcome depends. So, ultimately, you have to have hope for the human spirit, which, when called on, can reach great heights. But right now, if you compare the situation with what it was at the end of the Second World War, say, when I felt there was very considerable hope, you get discouraged. The catastrophe of the war had made people feel more responsible about the problems. Yet now, when the difficulties are greater, the men are lesser. You find great ingenuity exercised in the attainment of unworthy objectives.

"Of course, the individual everywhere is pretty decent. If he weren't, he probably would have been killed off when he was young. The number of actually wicked people is small. However, people may be decent, but they may be working in a civilization like ours, which sometimes drives people to the wall. It's a kind of rat race, where decency will kill you in certain circumstances. And the materialistic ideals in a country like ours make everything very difficult. How does a man achieve distinction? There are very few public honors. So he wants a bigger car and a house on the hill. If that's his aim—not even his own aim, but the expectation of people around him—he'll be as decent as he can up to a certain point, and then. . . . I don't really blame him. But I do blame a system of education that makes those things the ideal. Sometimes, though, it's not the education but what societal organization brings afterward. There are lawyers, for example, who in law school have the very highest ideals of the law and what it means to human organization, but after they are out and in competition they find that the life of a lawyer and the life of the law are two very different and sometimes opposite things. It is only in science, I find, that we can get outside ourselves. It's realistic, and to a great degree verifiable, and it has this tremendous stage on which it plays. I have the same feeling—to a certain degree—about some religious expressions, such as the opening verses of the Bible and the story of the Creation. But only to a certain degree. For me, the proper study of mankind is *science*, which also means that the proper study of mankind is man."

PART II

Three Faces of Biology

THE THREE PIECES of writing collected here deal with the experience of biologists. Biologists are now sailing in some pretty heavy water. I am not a biologist, but I do not take any satisfaction from the presently fashionable idea that biologists are beginning to know "sin"—because of recombinant DNA and genetic engineering—just as physicists knew "sin" because of the atomic bomb. I would argue, if this were the place to do it, that neither side of this equation makes much sense. What I am concerned with, rather, in these pieces are the kinds of difficulties that a society—the Soviet Union—can make for a science like biology and the kinds of difficulties that the biologists can make for themselves. (My profile of Lewis Thomas considers the kind of difficulties we *all* seem to be making for ourselves by our obsessive concern with our own health. What does it say about a society when it is the healthiest society—physically speaking—that has ever existed in human history, while, at the same time, it appears to be floundering in a kind of neurotic preoccupation with disease?) As a physicist I do not get any satisfaction out of the imbroglio over the credits due for the discovery of the structure of DNA. I have seen enough of this kind of thing in my own field to realize that this is an aspect of human nature and not something that is the special province of the biologists. Furthermore, as an American, I do not get any special satisfaction out of the barbaric disaster that was Soviet genetics in the 1930s. We should not forget that Scopes *lost* the "Monkey Trial" and that, consequently, Darwinian evolution was an illegal idea in the state of Tennessee until fairly recently. These pieces are cautionary. They are meant to chill, but not to freeze.

3

LYSENKO: ENEMIES OF THE PEOPLE

IN all the history of science there is no record of anything to equal what happened to Russian biology in the thirty years between 1934 and 1964. During this time—a time in which biology was making extraordinary advances—the whole science of modern biology was suppressed in Russia and vigorous efforts were made to erase it from the Russian mind. Textbooks were altered, biologists who were courageous enough to continue to try to teach and do research in real biology were persecuted or imprisoned. There were severe crises in Russian agriculture—crises that might have been alleviated by the use of modern biological techniques, among them the cultivation of hybrid varieties of corn and wheat, if such techniques had not been forbidden. In their place, an entire pseudo-science was erected and forced upon the Russian people practically at gunpoint by a fanatic, illiterate charlatan, Trofim Denisovich Lysenko, who had the complete confidence of both Stalin and Khrushchev. (It is likely that one of the key elements in the downfall of Khrushchev was the failure of Russian agriculture, and one of the first acts of the regime that replaced him was the denunciation of Lysenko and his followers, who held most of the important posts in biology and agriculture.) Much has been written about Lysenko, mainly by Western observers, but the best source is still the book by the distinguished Russian biologist Zhores A. Medvedev, *The Rise and Fall of T. D. Lysenko,* which appeared here in 1969. (It has not been published in Russia.) In its controlled fury, it has all the literary qualities of an Orwellian polemic; in its revelations of the anguish of men and women whose only crime was that they could not teach things they knew to be false, it often matches the great Russian novels; as an object lesson in the methods of dictatorships, it may be one of the definitive modern masterpieces.

Dr. Medvedev's book—he was at one time the director of the Laboratory of Molecular Radiobiology at the Institute of Medical Radiology, in Obninsk—was translated by I. Michael Lerner, a professor of genetics and once chairman of that department at the University of California at Berkeley. In a fascinating foreword to his translation, Lerner gives us the history of the book:

> Although I know only a few personal facts, I should like to recount the history of my connection with this book. In 1961 I received a copy of a book, in Russian, by Y. M. Olenov, for review in *Science.* It dealt with population genetics and evolution, and its main purpose seemed to be to present the developments in these areas to the Soviet scientific community which, under Lysenko's regime, knew nothing of them. It was a good book: Engels was mentioned in it only once, and the whole tenor was not one of demagogic style (described so vividly by Medvedev) but rather that of an objective and scientific spirit. My review was entitled "The Blossoms of a Hundred Flowers of Soviet Genetics," echoing the statement of Chairman Mao. In response to my review, a postcard came to me from the Laboratory of Radiobiology in Obninsk (a hundred and ten kilometers from Moscow), informing me that I was talking through my hat—for every flower there are still a hundred weeds, it said. The writer of the card, Medvedev, turned out to be a young man of high intelligence, spirit, and courage, a Soviet patriot, and an active participant in the struggle against Lysenkoism described in this book. . . .
>
> Medvedev and I struck up a correspondence and, at the Mendelian celebration in Czechoslovakia in 1965 . . . managed to meet each other. [The celebration was in honor of the hundredth anniversary of Gregor

Mendel's discovery of the laws of heredity; Mendel, who did his work in the gardens of a monastery courtyard in Brno, was a special target of Lysenko's propaganda, so Russian participation at this centennial was of particular importance.] He told me that he had been working since 1961 on a history of the whole sordid affair and showed me an outline of the book. I immediately volunteered to translate it when and if it was published in the U.S.S.R.

Subsequently Medvedev informed me that publication was to be delayed, because the powers that be had decreed that 1967, being the fiftieth anniversary of the Revolution, was not a suitable time to bring out books critical of the Soviet regime.

In the fall of 1967 I was instrumental in bringing to the United States a delegation of four Soviet geneticists (this was after Lysenko's fall), and at that time I discussed with them fully, frankly, and without reservations the prospects for doing the translation. To my utter amazement, at least some of the Soviet visitors assumed that I already had the manuscript in my possession and, as I found out later, one of them denounced Medvedev and me on their return. I have no first-hand information about what happened after the denunciation, but I do know that Medvedev must have been put in a highly embarrassing position, having been falsely accused of planning to pull a sort of scientific "Dr. Zhivago" or a Daniel-and-Sinyavsky attempt to publish his manuscript abroad without official sanction. He wrote suggesting that I request a copy of the manuscript from the publishing house of the Soviet Academy, with a view to translating it into English, and that I point out the obvious advantages of an authorized translation. . . . After several months of silence, I received a letter from the publishing house which indicated that the manuscript was not publishable in the U.S.S.R. and therefore I could not have a copy of it.

Meanwhile, through unofficial channels, I came into possession of a microfilm of the typescript. The author had circulated many copies of preliminary versions throughout the Soviet Union for the purpose of checking the accuracy of his account of the events described. The final Russian text, which provided the basis for the present translation, resulted from numerous revisions by the author, and has been approved by him as representing his current views. For obvious reasons, he did not see the translated, abridged, and edited manuscript before publication. It is hoped that he may one day see a copy of this book.*

T. D. Lysenko, who had been experimenting with a project for

*Medvedev emigrated to England in 1973.

planting peas as a cover for fields used as winter pastures, came to public notice in 1926. His experiments may or may not have had some value for agronomy, but at any rate they revealed some of the techniques that Lysenko was to make use of later in establishing his hold on Soviet science. Not contenting himself with the usual channels of scientific communication, he had managed to have himself made the feature of an article in *Pravda* entitled "The Fields in Winter." The author gave a graphic portrait:

> If one is to judge a man by first impression, Lysenko gives one the feeling of a toothache; God give him health, he has a dejected mien. Stingy of words and insignificant of face is he; all one remembers is his sullen look creeping along the earth as if, at very least, he were ready to do someone in. Only once did this barefoot scientist let a smile pass, and that was at mention of Poltava cherry dumplings with sugar and sour cream.

By 1929, Lysenko was well on his way with the "research" that ended in the destruction of the study of modern genetics in the Soviet Union. It began simply. Lysenko and his colleagues experimented to see what would happen to the seed of winter wheat if it were exposed to the cold and then planted in the spring. The experiments led Lysenko to the statement that the direct transformation of spring wheat into winter wheat, and vice versa, was possible—the process of "vernalization." His idea, which can be traced back to the pre-Darwinian French biologist J. B. de Lamarck, was that the adaptability of certain strains of spring wheat to winter conditions, and other characteristics, could be directly inherited by the offspring. This theory of "inheritance of acquired characteristics," which ran counter to every notion of modern genetics, became Lysenko's cause, and its acceptance, again at gunpoint, was forced upon Soviet biology. The issue is clear-cut. According to present-day biology, and, for that matter, to the biology universally accepted in the 1920s, heredity is controlled by the genes. A gene is a complex molecule whose structure determines the characteristics of the offspring of a given species. Such molecules can be altered (the process of mutation)—for example, by exposure to radiation or chemicals—and mutation produces abrupt and generally uncontrollable changes in the hereditary pattern. Mostly, these changes are "destructive"; the altered species that result cannot adapt to their environment and die out. Some mutations are beneficial, in that the altered species more readily adapts to its environment than its predecessor did, and this is how biological species "evolve." But the fact that a parent acquires a certain characteristic during its lifetime does *not* mean that the offspring will inherit this characteristic. An animal may lose a limb and a plant a leaf, but their descendants will be perfectly whole. To deny this, which in essence was what Lysenko was proposing, forces one to deny the whole of modern biology.

Sometimes a genius—an Einstein—can propose a theory that runs counter to all the accepted scientific dogma of his era and be right. This is a rare phenomenon. Mostly, the iconoclasts turn out to be cranks. However, early in the thirties, Lysenko convinced Stalin that his biology was the *only* one consistent with Marxist political ideology, and by 1935 Lysenko was calling scientists who did not accept his theories "saboteurs" and "enemies of the people." Medvedev notes that after a speech by Lysenko that ended with "a class enemy is always an enemy whether he is a scientist or not," Stalin exclaimed "Bravo, Comrade Lysenko, bravo!" By 1936, there were two biologies in the Soviet Union—the state biology of Lysenko, and a truly scientific biology practiced by a decreasing number of Soviet scientists who were prepared to go to prison, or even die, for what they knew to be the truth. Medvedev's book is filled with the names of Soviet scientists who simply would not give in to Lysenko, whatever the consequences. The most moving part of his account deals with the great Soviet biologist Nikolay Ivanovich Vavilov. Vavilov, a specialist in plant pathology, had set himself the task of developing for Soviet agriculture disease-resistant species of cultivated plants, and to this end he began the first broad study in the Soviet Union of applied plant genetics. Medvedev remarks:

> It was to further this work that Vavilov, in the middle twenties, initiated his famous expeditions to all corners of the Soviet Union and later to all principal centers of world agriculture. Over a short period of time about two hundred expeditions were organized. Their members investigated the agriculture and plant resources of sixty-five countries and brought to the Soviet Union over a hundred and fifty thousand plant varieties, forms, and species—all of the plant-breeding wealth created by mankind in its centuries-old history.

Vavilov, the most outstanding Russian geneticist, was instantly attacked by Lysenko and his followers, and in 1939 one of them wrote, "It so happened that, together with foreign plants, bourgeois theories and pseudo-scientific trends [i.e., modern genetic theory] infiltrated [Vavilov's] institute." Indeed, by this time Lysenko was trying to force Vavilov's resignation. Vavilov could not be silenced, and after one particularly acrimonious debate with him Lysenko remarked, "I say now that some kind of measures must be taken. . . . We shall have to depend on others, take another line, a line of administrative subordination." The "others" were, of course, the Soviet secret police. In August of 1940, Vavilov, on an expedition in the western Ukraine, was arrested. Medvedev describes Vavilov's last hours of freedom:

> Vavilov and his companions first went to Kiev. From there they went by car to Lvov and on to Chernovitsy. From there, in three overcrowded cars, Vavilov and a large group of local specialists proceeded toward the foothills to collect and study plants. One of the cars could not negociate the difficult road and turned back. On the way the occupants met a light car containing men in civilian clothes: "Where did Vavilov's car go?" asked one of them. "We need him urgently." "The road further on is not good, return with us to Chernovitsy. Vavilov should be back by 6 or 7 P.M., and that would be the fastest way to find him." "No, we must find him right away; a telegram came from Moscow; he is being recalled immediately."

In the evening the other members of the expedition returned without Vavilov. He was taken so fast that his things were left in one of the cars. But late at night three men in civilian clothes came to fetch them. One of the members of the expedition started sorting out the bags piled up in the corner of the room, looking for Vavilov's. When it was located it was found to contain a big sheaf of spelt, a half-wild local type of wheat collected by Vavilov. It was later discovered to be a brand-new species. Thus, on his last day of service to his country, August 6, 1940, Vavilov made his last botanical-geographic discovery. And although it was modest, it still cannot be dropped from the history of science. And few scientists reading of it in a Vavilov memorial volume published in 1960 could have guessed that the date of this find is a date that scientists throughout the world will always recall with bitterness and pain.

On July 9, 1941, after a brief "trial," Vavilov was convicted of "belonging to a rightist conspiracy, spying for England, leadership of the Labor Peasant Party, sabotage in agriculture, links with white emigrés," and so on. He was sentenced to death, but the wife of Beria, the police chief of Soviet Russia, was a student of plant breeding, and her professor, D. N. Pryanishnikov, persuaded her to persuade Beria to spare Vavilov. He was sent to prison, where, on January 26, 1943, he died of undernourishment. Medvedev reveals a significant fact about Soviet sensibilities:

> At the end of 1942, Vavilov, who had mysteriously "disappeared" from the world science scene, was elected a foreign member of the Royal Society of London. When this information reached the N.K.V.D., the Vavilov file was urgently recalled for study. But it was too late. Life was slowly ebbing from a body exhausted by malnutrition, and it was impossible by then to save him. . . . This was the heaviest loss to Soviet science in the period of the personality cult.

Vavilov provided his own epitaph when he wrote, "We shall go to the pyre, we shall burn, but we shall not renounce our convictions."

During the war, the persecution of biologists temporarily abated. But by 1945 Lysenko had resumed his activities with renewed vigor. Now Medvedev himself finally became aware of what Lysenko was about:

> Up to then, not really knowing genetics, I had viewed the controversy in genetics and Darwinism as a real scientific debate in which, as it appeared to me, both sides deserved respect. [One of the more bizarre aspects of the affair was the treatment of Darwin. The followers of Lysenko conceived the notion that Darwin's work, which coincided, according to them, with the last flourishing of the capitalist society, represented the last bit of valid biology done in the West. Hence there ensued an all but incredible debate, as to which group of Soviet biologists were the real representatives of the true "Darwinism."] But watching the renewal of the discussion on Darwinism, I understood that the aim of Lysenko and his followers was anything but elucidation of scientific truth. . . . It soon became apparent that the position of Lysenko and his followers was weak, far-fetched, and based on few facts. It really bordered on utter falsification of science. It also became clear that neither Lysenko nor his supporters were possessed of sufficient erudition to carry on the debate at the level of serious science.

In fact, by 1948 it appeared that Lysenko might be on his way out. However, once again, he went to Stalin, and a campaign was soon under way to examine every biology text and even the contents of every course in biology to make sure that it contained no mention of modern genetics; a biologist could be dismissed from his post if he even published a finding that happened to agree with a similar finding of a scientist in the West. Medvedev attended one of Lysenko's public lectures:

> An especially summoned brass band begins to play a triumphal march, under the sounds of which Lysenko proceeds through the hailing rows to the rostrum to begin his first lecture. Seeing gray-haired scientists in the front rows of the audience, Lysenko exclaims with exaltation: "Aha! You came to relearn?" I remember little of the content of the lecture—only the assertion that a horse is alive only in interaction with the environment; without interaction it is no longer a horse but a cadaver of a horse; that, when different birds are fed hairy caterpillars, cuckoos hatch from their eggs . . . etc. etc.

Whatever Stalin's death may have meant for other intellectual disciplines in the Soviet Union, it effected little change in the climate for biology. Khrushchev was a confirmed Lysenkoist, and until 1964, when he was deposed, true biology remained a forbidden, clandestine activity in the Soviet Union. It was only after 1964 that the full extent of the methods of Lysenkoism were revealed. (It turned out that the "data" Lysenko had gathered to support his theories about the collective farms were largely false, and that farmers using his methods had suffered such enormous losses that a crisis in Soviet agriculture ensued.) A desperate effort began to "rehabilitate" Russian biology. New textbooks were needed, and whole research establishments were created almost overnight. The day before Khrushchev was deposed, the noted geneticist I. A. Rapoport received a call from a high official in Soviet agriculture, which he assumed was to announce his dismissal from his job but which turned out to be a command to prepare, within twenty-four hours, a popular article for a Soviet newspaper on the achievements of modern genetics. Medvedev reports:

> Rapoport replied that he could not complete such a serious article so fast, and that in any case the newspaper, noted for its pogrom publications and Lysenko sympathies, would hardly be likely to publish it.

His caller informed Rapoport that he would receive all the secretarial help he needed and that the article would indeed be published. In thirty hours, Rapoport produced his article, and a week later it appeared, more or less as he had written it; what he had said in praise of Mendel was, however, deleted.

Medvedev does not really explain how men like Rapoport, or indeed he himself, were able to keep abreast of modern genetics during the Lysenko regime. One can only imagine the many acts of courage that this must have required. The damage Lysenkoism caused will be felt in Soviet science and agriculture for decades. Medvedev does indicate that Lysenkoism has been largely eliminated from Soviet biology, but it is evident that the struggle for intellectual freedom in Russia is as anguished as ever. This does not mean that we should indulge in self-congratulation. Every country has its potential Lysenkos, entirely too willing to convert intellectual debates over controversial and difficult issues into denunciations of the "enemies of the people." We can only hope that if the need arises, we, too, can produce, in large numbers, people with the character, the intellectual honesty, and the moral courage of Medvedev to carry on the struggle for truth.

4

A SORROW AND A PITY: ROSALIND FRANKLIN AND THE DOUBLE HELIX

IN APRIL of 1968, I had the opportunity of reviewing, for *The New Yorker, The Double Helix* by James D. Watson, a book I found fascinating. It was clearly not a formal history of the discovery of DNA but a personal attempt by Watson to recreate events as he had seen them some fifteen years earlier. In my review, I quoted the following passage from the book's introduction, which was written by Sir William Lawrence Bragg, who was the director of the Cavendish Laboratory in Cambridge, where Watson and Francis Crick did their work. Bragg wrote: "Finally there is the human interest of the story—the impression made by Europe and by England in particular upon a young man from the States. He writes with a Pepys-like frankness. Those who figure in it [Bragg was one] must read it in a very forgiving spirit. One must remember that his book is not a history but an autobiographical contribution to the history which will someday be written." So it seemed to me as well.

Soon after this review was published, I received a letter from a woman—Anne Sayre—whom I did not know. (This may be a good time, in view of what will follow, to state that I do not know *any* of the principals in this matter, and in that sense I do not have any personal axes to grind.) Mrs. Sayre informed me that while she herself was not a scientist she had been a close friend of Rosalind Franklin, one of the important figures in Watson's narrative. Rosalind Franklin died of cancer at the age of thirty-seven, in 1958. Mrs. Sayre had been profoundly outraged by Watson's characterization of her late friend, whom Watson consistently referred to as "Rosy"—a sobriquet, Mrs. Sayre noted, that was never used by any of Miss Franklin's close friends. (In fact, Watson added an epilogue to *The Double Helix,* a sort of apologia for his treatment of Miss Franklin. This was one of several cases where, in his attempt to recover the events that actually occurred between 1951 and 1953, Watson made no attempt to try to make himself look better than he thought he had been.) Mrs. Sayre wrote in her letter to me that she felt Watson's treatment distorted not only the character and personality of Rosalind Franklin but her contributions as well.

I was very troubled by this letter, but since I had no firsthand knowledge of any of the events and people involved, I was in no position to evaluate it. I felt this was something that would get sorted out once the professional historians of science began to pour over the vast assortment of documents—letters, laboratory reports, published and unpublished papers, and even tape-recorded interviews—that were available. This process is now well underway, and a certain milestone was reached with the publication, in 1974, of an extraordinary history of the whole subject written by Robert Olby and entitled *The Path to the Double Helix.* Olby is a lecturer in philosophy at the University of Leeds, and from his book one gathers that he is a highly qualified scientist as well. I will return to a brief general discussion of Olby's book, but, for the moment, let me quote a characterization written by Francis Crick in the forward to the book: "What I think certain is that no future historian of science in this area will be able to ignore the present volume, both for the thoroughness of Olby's investigation and for the good judgment he brings to the task."

Mrs. Sayre, however, not content to wait for the historical process, has taken the matter into her own hands and has written a book entitled *Rosalind Franklin & DNA*. Several favorable reviews of this book appeared, and it became something of a feminist *cause célèbre*. (On the front cover the book is described as "a vivid view of what it is like to be a gifted woman in an especially male profession.") What concerns me is not this, but whether the account on which all of these conclusions have been based is free from significant misinterpretations of fact—misinterpretations that might reflect on the conclusions themselves. In what follows, I will attempt to show that there are such misinterpretations, and that, at least for me, they cast doubt on any conclusions I would otherwise draw from this book.

At the outset I would like to state my ground rules. My only sources have been Mrs. Sayre and Olby. In her list of acknowledgments, Mrs. Sayre writes, "Dr. Robert Olby has generously shared some of the material from his forthcoming book on the history of the discovery of the structure of DNA." Perhaps she did not see all of Olby's book, although it was published about a year before her own. This is the only generous interpretation I can find for what has occurred.

I will avoid material based on interviews given to Olby as early as 1968, since these were given some fifteen years after the events in question occurred. I will restrict myself to only those matters that are documented by contemporary letters, papers, and laboratory reports. Where it is relevant, I will reproduce these in full so that the reader may form his or her own judgments. Mrs. Sayre has also done extensive interviews with the same people, and I wish to avoid entirely the question of the consistency of these various interviews. Anyone who has engaged in this kind of interviewing will know that the same individual can say different things on different days to the same interviewer, let alone to different interviewers. I have had the experience of faithfully reproducing such a tape-recorded interview only to find that my subject denied ever having said any such thing. I simply wish to avoid this here.

Let me begin on a positive note by describing some aspects of Rosalind Franklin's life and career about which there appear to be no disagreements. Rosalind Franklin was born in London on July 25, 1920. Her family was an old British-Jewish family of great distinction and considerable prosperity known for their work in charitable and liberal social causes. Her mother was a Waley, and she was related to the great poet and Orientalist Arthur Waley. Rosalind Franklin had a private income throughout her life which she rarely used. She did not have to take up a career to support herself, and so her choice of science was an act of pure dedication. She was an absolutely dedicated scientist, and a truly first-rate one. Her father wanted her to go into some sort of social work, but over his objections she decided, at Cambridge, to devote herself to science. I will resume an outline of her career shortly, but here I wish to give some impressions of her personality; all of these I have taken from Mrs. Sayre's book. Indeed, one aspect of her book that I admired was her attempt to give a rounded picture of Rosalind Franklin which includes a description of the "difficult" side of her character. I will begin with comments related to this side of her character—the only side that Watson noticed during the years from 1951 to 1953. His portrait is two-dimensional, but I want to include these assessments to show that he was describing *something.* That he and his colleagues missed the rest is one of the real tragedies in this story. Here, then, are several observations taken from Mrs. Sayre.

Rosalind Franklin's mother noted that, "Rosalind felt passionately about many things, and on occasion could be tempestuous. Her affections both in childhood, and in later life, were deep and strong and lasting, but she could never be demonstrative, or readily express her feelings in words. This combination of strong feeling, sensibility and emotional reserve, often complicated by intense concentration on the matter of the moment . . . could provoke either stony silence or a storm . . . the strong will and a certain imperiousness and tempestuousness of temper, remained characteristic all her life."

Rosalind Franklin's thesis adviser at Cambridge saw her as "stubborn and difficult to supervise" and as "not easy to collaborate with." Maurice Wilkins, her colleague at King's College during the period of the DNA research, saw her as "very fierce, you know. She denounced, and this made it quite impossible as far as I was concerned to have a civil conversation. I simply had to walk away." Her student and collaborator Raymond Gosling said: "She didn't suffer fools gladly at all. You either had to be on the ball, or you were lost in any discussion about anything, and that was constant." Finally, there is Mrs. Sayre: "Rosalind's view of marriage remained until she was in her thirties completely based upon what she had seen of her parents' marriage, which she took literally for a model and an invariable pattern: strong dominant male, supportive female, and a home in which the wife, especially in her role as mother, was firmly, permanently and wholly centered. This was so rooted in her that, indeed, no man who was not very strong in the way in which her father was strong ever attracted her. More than that, for men who were weak, submissive, pliable, unserious, unforthright, she had a lack of respect which at times came close to contempt."

On the other hand, Mrs. Sayre found in her a deep shyness—an apposition of character traits that is not unfamiliar, but that can wreak havoc in personal relationships. The great pity is that no one was available during Rosalind Franklin's tenure at King's College to break through her isolation. This may have been related, as Mrs. Sayre believes, to the fact that she was the only woman in a male-dominated department; and it also may have been related to the generally admirable British characteristic of minding one's own business. There are times when it is helpful to have someone around who at least understands what various people's emotional business *is*, so as to try to bring people together. If this had happened in the period from 1951 to 1953, the history of the discovery of DNA, as I will indicate shortly, might have been very different. To the few people who got to know her, including Mrs. Sayre, she was, apparently, a wonderful friend, both kind and humorous. Also, and this is something that one would never gather from Watson's portrait, she was a very handsome woman. Mrs. Sayre has included a photograph of her—taken, one gathers, on a skiing vacation—that is truly stunning. In this respect, Watson's book misled me completely.

Upon leaving Cambridge in 1942, Rosalind Franklin, now trained as a physical chemist, took a position with the British Coal Utilization Research Association (BCURA), an organization designed to study the physics and chemistry of coal. Her work there is outside the domain of this essay, except to quote an evaluation of it by Professor Peter Hirsch of Oxford. It was, he wrote, "remarkable. She brought order into a field which had previously been chaos." She remained with BCURA until 1947, when she went to Paris; she remained there until the end of 1950. In Paris, she was appointed a *chercheur* in the *Laboratoire Central des Services Chimiques de l'Etat*. By all accounts, the three years she spent in Paris were the happiest of her life. She made lasting friendships, went skiing and mountain climbing, and also, changed her field of research. She became an X-ray crystallographer (crystallography is the study of the structure of crystals by X-ray methods). But by 1949 she felt that she should reestablish herself in England, and so she applied for a fellowship. She knew nothing about biology, but she wanted to begin to study, by X-ray techniques, biologically important molecules. She wrote in 1950, to a senior colleague in Britain, "I am, of course, most ignorant about all things biological, but I imagine most X-ray people start that way. . . ." She hesitated between King's College and Birkbeck, both in London, and finally chose King's. In late 1950, Rosalind Franklin arrived at King's, and it is just at this point that the problems in Mrs. Sayre's account begin.

Rosalind Franklin's difficulties at King's started almost at once, and involved the precise conditions of her appointment. Much of her subsequent difficulties can be traced to this. Mrs. Sayre writes, "When Randall [This is J.T. Randall, later Sir J. T. Randall, the director of the laboratory at King's where Miss Franklin worked. This is the same J. T. Randall who was one of the co-inventors of the magnetron vital for radar.] offered Rosalind a Turner-Newell Research Fellowship it was on the understanding that she would be put in charge of building up an X-ray diffraction unit within the laboratory, which at that time lacked one, and this is what she set about doing. She had not been brought to King's College to work upon DNA or any other specific problem. . . ." Here, on the other hand, is a letter, quoted in Olby, written by Randall to Miss Franklin just prior to her arrival at King's. The letter is dated December 4, 1950:

> After very careful consideration and discussion with the senior people concerned, it now seems to us that it would be a good deal more important for you to investigate the structure of certain biological fibres in which we are interested, both by low and high angle diffraction, rather than to continue with the original project of work on solutions as the major one.
>
> . . . as far as the experimental X-ray effort is concerned there will be at the moment only yourself and Gosling, together with the temporary assistance of a graduate from Syracuse, Mrs. Heller. Gosling, working in conjunction with Wilkins, has already found that fibres of desoxyribose nucleic acid [DNA] derived from material provided by Professor Signer of Bern gives remarkably good fibre diagrams. The fibres are strongly negatively birefringent and become positive on stretching, and are reversible in a moist atmosphere. As you no doubt know, nucleic acid is an extremely important constituent of cells and it seems to us that it would be very valuable if this could be followed up in detail. . . .

I beg the reader's indulgence for having quoted parts of this somewhat technical letter. However, from it it is very clear that Miss Franklin *was* brought to King's College to work on DNA. It is also clear that there existed, prior to her arrival, a serious program at King's, one that was already well-established. I will return to this in a moment. Now Mrs. Sayre, having gotten Miss Franklin to King's with no problem to work on, proceeds to "explain" how Miss Franklin got into working on DNA. She writes, "The available problems were numerous, but of all the projects underway in the biophysics department of King's College in 1951, the one devoted to the investigation of DNA was not only the most important, but to any imaginative scientist, the most provocative and fascinating. Rosalind could not resist it, and for good reason. . . ." Mrs. Sayre then presents a brief history of genetics, ending with the following quotation from *Molecular Genetics: An Introductory Narrative,* written in 1971 by Gunther S. Stent: "The way was now clear to formulate a theory of how DNA can act as the carrier of genetic information. . . . It seems impossible today to say who was actually responsible for originating these notions. The theory suddenly seemed to be in the air after 1950 and had come to be embraced as a dogmatic belief by many molecular geneticists by 1952. The key proposition of this theory is that if the DNA molecule contains genetic information, then that information cannot be carried in any way other than as the *specific sequence* of the *four nucleotide bases* along the polynucleotide chain." Mrs. Sayre then comments, "This, then, may be taken as the general state of affairs with respect to DNA when Rosalind went to King's College. The problem was irresistibly attractive. . . ."

In fact, when Miss Franklin went to King's College in 1951, next to nothing was known about the role of DNA in genetic heredity, or about its structure. The last section of the quote from Stent refers to the "central dogma" of the DNA theory of genetic replication, which was formulated after Watson and Crick found their model *in 1953.* None of this, of course, was known to anyone at King's College early in 1951, least of all to Miss Franklin, who was, as she herself had said, "most ignorant of all things biological." From Randall's letter, it is clear that Miss Franklin began working on DNA because she was asked to.

In fact, what did Miss Franklin inherit when she came to King's? There was Maurice Wilkins's graduate student Gosling. It was Gosling who had actually taken the X-ray photographs of the DNA mentioned in Randall's letter. There was also the lab's supply of DNA. This was not a trivial matter. Very little DNA, which is extracted from living cells, was available in the world in 1950. It happened that Professor Rudolf Signer of Bern had come to a meeting in England on May 12 of that year with a bottle of his best DNA, which he distributed, and Wilkins got a sample. It was after this that the research began. Finally, there was what I call the "helical gestalt." Because of Linus Pauling's successful work on helical protein structure in 1948, it was natural for these early DNA researchers to take as a working hypothesis that DNA also had some sort of helical structure. Indeed, by the summer of 1951 Maurice Wilkins was already considering a *single* helix model for DNA, which he thought fitted his X-ray data taken in 1950. (We now know, to be sure, that it is a *double* helix—like a circular stairway.) Indeed, at his urging, A. R. Stokes, at King's, had worked out the mathematical theory of X-ray diffraction from a crystal of helical molecules, a bit of careful mathematics that was also done independently by Crick and W. Cochran at Cambridge. This had been done in the summer of 1951, as mentioned above, and Miss Franklin began taking her own X-ray photographs only in September of 1951. She was well aware of Wilkins's helical interpretation of his data because at about that time, when Wilkins was in the United States in search of more DNA, he wrote to her, as quoted by Olby: "The structure might be a 40° pitch single helix, one per cell, the layer lines being given by the pitch of the helix and the nucleotides uniformly spaced along the helix. . . ."

I mention this because here is how Mrs. Sayre characterizes the work of Wilkins, Gosling, and Stokes: "Before she [Franklin] arrived some attempts had been made at X-ray diffraction photographs of DNA (recent attempts that is: W. T. Astbury had made photographs much earlier) [1938-39] without very promising results, which is not surprising in view of both the lack of experience in the experimenters and the simplicity of the equipment. . . ." One of the most totally unacceptable aspects of her book, of which the above paragraph is an example, is the way in which Mrs. Sayre denigrates the research of the other scientists who were involved in the study of DNA. At times her prose becomes almost Nixonian; for example, "Plainly Watson admired Crick from the start, and certainly their alliance was a sound and productive one. Though one rather uncharitable commentator upon the events surrounding the establishment of the structure of DNA is given to describing Watson as 'a leg man for Crick,' this view will not stand up to analysis. It is too serious an underestimation both of Watson's abilities, which are far from mean, and of what Crick can accomplish on his own without such devices." Who was this "uncharitable commentator"? If this preposterous characterization "will not stand up to analysis," then why is it here at all? And why is it followed by a condescending but authoritative—albeit grudging sounding—denial by Mrs. Sayre? One must bear in mind that Mrs. Sayre is not a scientist. Yet, in this book, she pontificates like some sort of supernumerary Nobel Prize committee. There is a maxim of Wittgenstein's—one of the few of Wittgenstein's maxims that I claim to understand fully—which states *"Wovon man nicht sprechen kann, darüber muss man schweigen"* ("Whereof one cannot speak, thereof one must be silent").

Soon after Miss Franklin began doing her X-ray work, Wilkins returned from America with a new sample of DNA. What he found on his return is described by Olby as follows: "All was well while Wilkins kept to his optical studies or attended conferences. But when he returned from Chargaff's laboratory [Professor E. Chargaff, of Columbia University, was one of the early pioneers in the study of DNA; it was from him that Wilkins had obtained a second sample of DNA] and expected to join in the X-ray work Franklin was outraged, and matters between them were only temporarily smoothed over by Wilkins agreeing to leave the Signer DNA to Franklin and Gosling whilst he used the Chargaff DNA." From this time on—though Wilkins and Miss Franklin were doing experiments on DNA in the same basement laboratory complex at King's—there was nothing less than open warfare between them. It lasted from 1951 until 1953, when Miss Franklin left King's for Birkbeck College. The worst aspect of this from the scientific point of view—even from their own point of view—is that if they *had* been able to arrive at some sort of working interchange of ideas, it is quite possible, even likely, that between them they would have found the structure of DNA before 1953.

Once Miss Franklin was able to begin her own X-ray work, she made rapid progress—at least for a while. Her work can be divided into three stages. This is important to understand, because Mrs. Sayre had also misinterpreted this aspect of the story. First, Franklin improved upon a technique that had been initiated by Wilkins for "hydrating" the DNA—making it take up water, which it did not want to do. Wilkins and Gosling had first X-rayed the crystalline form of extracted DNA—this became known as the A-form. It was this X-ray work that had convinced Watson, when he saw it, that DNA had an interesting crystal structure. Watson had been worried that molecular DNA might turn out to be some sort of amorphous mess. Apparently, he was unfamiliar with the prewar X-ray pictures of DNA taken by Astbury and his colleague Florence Bell; these, in fact, showed evidence of an interesting crystalline structure. Miss Franklin and Gosling found that the DNA, when hydrated, changed form dramatically—a transition from the A-form to a so-called B-form was induced. She and Gosling X-rayed the B-form and produced the best X-ray photograph of DNA that had ever been taken. (These photographs are reproduced in Olby's book.) Franklin correctly interpreted the photographs as evidence of DNA's helical structure, and was able to rule out a one-strand helix in favor of, and here I quote from laboratory notes cited by Mrs. Sayre (Olby uses a somewhat different version of these notes, which he admits he has edited): "a helical structure (which must be very closely packed) containing probably 2, 3 or 4 co-axial nucleic acid chains per helical unit, and *having the phosphate groups near the outside*. . . ." The italicized phrase is very important, because Watson and Crick struggled for two years trying to make models in which the phosphate groups were on the *inside* of the helix. The irony is that Watson attended a lecture given by Franklin in 1951 at which this matter was discussed. Watson took no notes, and by his own admission did not fully understand the lecture. If he had, he and Crick might have gotten their correct model much earlier. I will return to the matter of how Watson and Crick finally acquired access to Franklin's data, two years later, since it is one of the most controversial aspects of this tale. But it is important to point out—I cannot find any mention of this in Mrs. Sayre's book—that Franklin and Gosling never published any of this work in the open literature until 1953, something that I find incredible. They simply treated it as private and personal data. (The person who eventually suffered most from this was Linus Pauling, since Watson and Crick finally did get to see the data. Pauling became interested in the structure of DNA in 1952, but had only the prewar X-ray photographs and some inferior ones taken at the California Institute of Technology to work with. He was scheduled to come to London in the summer of 1952, when the U.S. State Department, for some reason understood best by itself, revoked his passport. Since Pauling made a valiant attempt at the structure problem, even with the old data, one asks what would have happened if Franklin and Gosling had published their new data in the open literature, where he could have studied it.)

Late in 1951, Franklin and Gosling switched their attention back to the A-form of DNA—the crystalline form—since the X-ray photographs showed a fascinating complex pattern. By February of 1952, Franklin was convinced that these X-ray photographs also showed a helical structure. But by May of 1952, she had changed her mind, and it is this change of mind that Mrs. Sayre simply refuses to recognize. She writes, for example: "Rosalind's success with the A-form has not been much commented upon. In the end the B-form proved more productive [This, by the way, is a bit moot. When Crick finally did get to see her A-form data, he was able to draw the conclusion—an incredible bit of mathematical deduction—that the two strands of the helix had to be running anti-parallel to each other. This fact had been buried in the data, and had gone completely unrecognized by Franklin and Gosling.] in suggesting a structure—a fact which, it ought to be said, Rosalind also came to realize [in January of 1953]—and consequently her work upon the A-form has been mostly ignored, except where it had been useful for the purpose of accusing Rosalind of being, curiously, fanatically, and immovably 'anti-helical' in her estimation of the probable structure of DNA. What is truly curious is that there is not a grain of justification for the accusation."

But here is a quotation from Miss Franklin's notebook for July of 1952, given in Olby: "There is no indication of a helix of diameter 11° A. The central banana-shaped peak fits curve calculated for helix of diameter 13.5° A having two turns/unit cell. If a helix there is only one strand. . . ." Olby writes, "according to Crick and Klug [this is A. Klug, one of Miss Franklin's last collaborators after she left King's] Franklin was considering three separate helices of diameter 11° A running side by side . . . any one such helix should give phosphate vectors at 11° A. Instead Franklin found it was at 13½° A. Therefore she rejected helices. Now for some reason—anti-helical prejudice, or just a natural failure to entertain the idea—she did not pursue *other* helical possibilities. For instance, she mentioned the possibility of two co-axial chains (a double helix) but did not realize, incredible as it may now appear, that such a structure would account for the peak at c=14° A in terms of halving of the 28° A repeat.

. . ." Again, I beg the reader's indulgence for this technical jargon. The point, I think, is clear, and is even clearer by Olby's summary remarks: "I would . . . suggest that from May, 1952 until Watson's visit in January, 1953 Franklin would gladly have buried helical DNA for the B-form as well as for the A-form. This would account for the mock funeral of the helix which was jokingly held at King's in 1952, for the skull and crossbones besides Gosling's conclusion, from a model building, that DNA must be a helix (undated manuscript but probably February 1953), and for the anti-helical impression Franklin gave to her colleagues and to Crick, who, in the summer of 1952, assured her, to no avail, that the double-orientation photo was misleading. Did she revert to her February, 1952 position when she heard about Pauling's helical model? It could well be."

Now, Mrs. Sayre is fully entitled to her notion of what constitutes a "grain of justification," but I am convinced that Miss Franklin did indeed change her mind about helices, which is no disgrace, following a blind alley until January of 1953—it happens to the best of us. She was a good enough scientist to get herself untracked at that time, but it was too late, since by then Watson and Crick had essentially the double-helix structure. This brings us to the penultimate matter I wish to discuss in connection with Mrs. Sayre's book. I am saving for last a discussion of her conclusions about all of this. It is the question of the degree of propriety shown by Watson and Crick in using, and acknowledging the use of, Miss Franklin's data—it is something that has been widely commented on. In the last analysis, what is involved are matters of ethical judgment which, in science, as elsewhere, are fairly subjective. I would like to enumerate the contacts that Watson and Crick had with Miss Franklin's data and explain, as best I can, how these came about. Before doing so I want to emphasize that there is no question of their having published "pirated data." There are no data at all in the famous Watson-Crick paper that was published in 1953. The data were presented in two accompanying papers, one by Wilkins and his group and the other by Franklin and Gosling. The question is one of access to this data prior to publication. Here is how access seems to have come about.

Watson attended the colloquium given by Miss Franklin (Wilkins and Stokes also spoke) on November 21, 1951. Whatever was presented at this colloquium was, as is customary, available for use by all participants—with, of course, proper citation. Much of the data that Watson and Crick eventually needed were presented at the colloquium, but Watson failed to take notes and did not fully appreciate what he had heard.

Within a week of this lecture, Watson and Crick made their first attempt at a model of DNA—it was a three-strand model. The group from King's, including Wilkins and Franklin, was invited to Cambridge for the unveiling. It was a disaster, since the model contradicted Miss Franklin's data, which had been misunderstood by Watson. At this unhappy juncture—and this is not even mentioned by Mrs. Sayre—Watson and Crick made an offer of collaboration with the King's group. Olby comments: "Watson and Crick tried to save the day by suggesting a future course of action in which the two groups would collaborate. But Franklin and Gosling, very understandably, would have nothing to do with such a suggestion. They had witnessed two clowns up to pranks. Why should they condone their behaviour by joining forces with them? And so back went the four King's scientists to London, leaving Watson and Crick deflated in Cambridge." We must bear this in mind when we examine Mrs. Sayre's conclusions.

Now comes the most serious matter. In December of 1952, Randall hosted a meeting of the Biophysics Committee of the British Medical Research Council at King's. It is likely that the purpose of the meeting, which was attended by very distinguished senior scientists, was to get an overview of the work being done in the field. In any event, Randall had prepared a report on the work being done at King's, and part of this report was written by Franklin and Gosling. It contained a summary of their work on

DNA, with the X-ray photographs appended. What happened next is described by Max Perutz, a very distinguished chemist from the Cavendish Laboratories at Cambridge who won the Nobel Prize in chemistry the year that Crick, Watson, and Wilkins won theirs in medicine (1962). Perutz had been at the meeting. He wrote in a later account in *Science* quoted by Mrs. Sayre, "As far as I can remember, Crick heard about the existence of the report from Wilkins, with whom he had frequent contact, and either he or Watson asked me if they could see it. I realized later that, as a matter of courtesy, I should have asked Randall for permission to show it to Watson and Crick, but in 1953 I was inexperienced and casual in administrative matters and since the report was not confidential, I saw no reason for withholding it." Without the use of the data contained in this report, it is doubtful that Watson and Crick would have found their successful model when they did.

I will return in a moment to the question of data obtained from Wilkins. This is a somewhat separate matter, although, as we shall see, there is a slight overlap. All of this, however, was acknowledged in the Watson-Crick paper by the remark, "We have also been stimulated by a knowledge of the general nature of the unpublished experimental results and ideas of Dr. M. H. F. Wilkins, Dr. R. E. Franklin and their co-workers at King's College, London." This acknowledgment, in my view, is adequate, but "close." In the best of all possible worlds, they might have added the details of what documents and colloquia these unpublished results were taken from, and they might even have thanked Perutz for allowing them to see the report. But in the best of all possible worlds, there never would have been the necessity for any of this. The people at King's would have published in the open literature, and various collaborations would have taken place.

The matter with Wilkins troubles me because I do not think that Watson and Crick treated him entirely fairly. This, of course, has nothing to do with Rosalind Franklin, but it is discussed by Mrs. Sayre, so it should be discussed here. After the disastrous visit to Cambridge by the King's group in 1951, the director of the Cavendish Laboratory, Bragg, simply ordered Watson and Crick to stop working on DNA. One of the strangest aspects of this whole affair is that laboratory directors in England, at this time, seem to have been in the habit of telling their charges simply to stop working on certain problems if this work "interfered" with what was being done at some other institution. It seems incredible, but that is how it was. Hence, for a full year, from late 1951 to January of 1952, the King's people had the DNA problem entirely to themselves. It was the appearance in England, in January of 1953, of the early versions of a paper by Pauling on DNA that opened up the field. Even though the paper was wrong—something discovered by Watson—its appearance made it clear that the DNA structure problem was now going to be worked on by everyone, and that no laboratory director had the slightest chance of stopping anyone from working on it. (However, Miss Franklin, when she left King's for Birbeck a few months laMer, was instructed, according to Mrs. Sayre, to stop even *thinking* about DNA. In fact, Franklin never did any further work on DNA when she left King's, but changed her field to the structure of viruses, a field in which she also made distinguished contributions right up to the time of her death.) Hence, in January of 1953, Watson and Crick resumed their active work on the DNA model, but they failed to inform Wilkins that they had done so. When, in January, Watson went to visit Wilkins at King's, in search of data, he should have, in my view, informed Wilkins that he and Crick were back in business. At this point, Wilkins would have been within his rights to limit his discussion of the data, since he had, by now, duplicated Rosalind Franklin's work and was actively in pursuit of the correct model himself. Whatever Wilkins's disappointment may have been at being beaten, he wrote to Watson and Crick after seeing a prepublication copy of their paper in March of 1953. (He also approved the form of the acknowledgment of his work.) "I think you're a couple of rogues but you may well have something. I like the idea . . . as I was back again on helical schemes I might, given a little time, have got it. But there is *no good grousing*—I think it is a very exciting notion and who the hell got it isn't what matters. . . ."

Finally, in view of all of this, I want to turn to Mrs. Sayre's conclusions. Her position is that Rosalind Franklin has been subject to a "slow and gentle robbery" of her ideas and credit—a robbery that began when Watson robbed her of her real personality in *The Double Helix*. I agree with her only to the extent that anyone who takes Watson's book seriously as history will come away with a distorted view of Rosalind Franklin and her work. It takes a very sophisticated reader of Watson's book to see—and this even Mrs. Sayre acknowledges—that a careful reading of it does give credit to Franklin for what she really did. I have, following Olby, tried to present a reasonably complete and balanced picture of this work—something that Mrs. Sayre, in her eagerness to obtain justice for her dead friend, has not, I think, done. This is what Mrs. Sayre thinks Watson and Crick should have done in 1953. She writes: "A wholly original paper on the base pairing scheme might have been written and published wholly by Crick and Watson, and a brilliant and insightful paper it would have been, too, quite enough of both to insure them lasting fame. A joint paper embodying the whole structure might have been written in which the contributions of Crick and Watson and of Rosalind Franklin would have been accurately and wholly defined, and though the glory attached to discovering the structure would then have been somewhat more divided, history would never have been confused concerning precisely who-did-what. There was, indeed, glory enough to go around.

"Neither of these alternatives was adopted. To be honest, the glory was hogged. . . ."

I have to make several comments about this idea:

Since Rosalind Franklin had refused Watson and Crick's collaboration in 1951, when they needed it the most, why should they have offered her a collaboration in 1953 *after* they had found the structure? As I have tried to indicate, it is not simply "base pairing" or any one single idea that Crick and Watson contributed to solving the structure of DNA. They made the double-helix scheme work. It is as simple as that. And, in her pursuit of justice, how has Mrs. Sayre accounted for Wilkins, let alone Gosling, who was Miss Franklin's collaborator on every paper she wrote on DNA? Mrs. Sayre's idea of glory-hogging seems very limited indeed.

Finally, let me comment on the idea of *post-factum* scientific collaboration by telling a brief anecdote involving myself. Some years ago, I was involved in a collaboration with two very distinguished senior physicists. In due course—something like a day and a half, in this case, since my colleagues worked with incredible speed while I chugged along behind as best I could—a preliminary version of a paper was produced. At this time, there was an important meeting of the American Physical Society in New York. Clutching our paper, I went to the meeting, where, to my elation, the first person I met was another senior colleague, one of the world's greatest experts on the very subject about which we had written our paper. I showed him the draft and was horrified when he informed me that, to his certain knowledge, our results had already been obtained by yet another distinguished physicist in a far-off city. With great unhappiness and trepidation, I reported this conversation to my senior colleagues. They were very surprised, since they knew the literature well and were not aware of any such work. They engaged in a search of the literature, which turned up nothing—but this was not surprising, since nothing had been published. At this point their moral obligation was at an end. (To a working scientist, what has not been published must be assumed not to exist. Part of the job of a serious scientist is to publish what he believes in. If he fails to do this and someone later rediscovers his work, one may admire the first man for his ingenuity, but, in most cases, he is out of luck.) However, my colleagues did not stop after searching the literature. They did something the moral character of which I have never forgotten, something which, according to the rules of the game we play, they were not required to do. They called the physicist in question and asked him if he had considered our problem. It turned out that he had, and was indeed preparing a long manuscript, which he sent us. There was a considerable degree of overlap between our work and his, and we agreed on a joint publication. Here, "overlap" is the key idea. In view of what has gone before, the reader may decide whether a joint publication of the DNA work would have been fair to anyone. I do not think so.

Now for a few final words about Olby's book, which I strongly recommend to anyone who wants to look more deeply into this subject. It is a very difficult book for a non-scientist to read. This may be characteristic of all profound books in the history of science—and this book is one of them. It sweeps over the whole history of events leading to the discovery of DNA, events that took at least half a century and culminated with the double helix. The care with which it is written is remarkable; there are over fifty pages of references alone. The perspective is vast, and the episode of the final discovery of the double helix is reserved for last. Rosalind Franklin finds her just place here—no more and no less. If Mrs. Sayre worried about Rosalind Franklin's place in history, she need not have. The matter is in firmer hands than her own.

5

LEWIS THOMAS: LIFE OF A BIOLOGY WATCHER

Until the spring of 1974, Dr. Lewis Thomas was known widely in the biomedical profession but hardly at all outside it. Dr. Thomas is president and chief executive officer of the Memorial Sloan-Kettering Cancer Center, in New York, which is one of the most important hospitals and cancer-research laboratories in the world, and since 1973 he has also been a professor of medicine and pathology at the Cornell University Medical College and since 1975 an adjunct professor at The Rockefeller University. For thirty-five years, he has done outstanding research in biomedicine—work for which he was elected to the National Academy of Sciences in 1972 and to the American Academy of Arts and Sciences in 1961. By the time of his election to the National Academy, Dr. Thomas had written some 200 technical papers on immunology, experimental pathology, and infectious disease—and had done so despite the fact that since 1954 he has been able to do laboratory work only part-time because of a heavy schedule of activities as a medical administrator. At present, he is, among other things, a trustee of the Rockefeller University and the Guggenheim Foundation, and last fall he was elected to the Board of Overseers of Harvard College, from whose medical school he graduated cum laude in 1937.

In the spring of 1974, Thomas published a collection of essays that had first appeared in the weekly *New England Journal of Medicine,* for which since 1971 he has been writing a monthly column of biomedical and philosophical commentary under the general rubric "Notes of a Biology Watcher." He called the collection, which was published by Viking, *The Lives of a Cell.* Despite the fact that the book is somewhat recondite reading—there is no technical glossary to explain such terms as "chronic glomerulonephritis" and "the cilia of eukaryotic cells," and Thomas assumes knowledge of these and other technical concepts—it has been a popular success. By October 1977, it had gone through ten printings in hardcover, six printings in paperback, and several book-club editions. In the United States alone, it has now sold more than 200,000 copies in hardcover and paperback, and it has been translated into nine languages, an incredible circulation for a book of this type. The response to the book from both literary and scientific readers was overwhelmingly enthusiastic, and in the spring of 1975 Dr. Thomas received a National Book Award for *Lives.* So unusual was the book that it created something of an embarrassment for the judges. Both the arts-and-letters panel and the panel on the sciences nominated Dr. Thomas, and an amicable tug-of-war ensued between the two; in the end, the arts-and-letters people conferred the award.

What particularly has struck many readers of the book was the combination of its language and its outlook. As for its language, describing a group of New Yorkers watching a collection of two million Central American army ants that had been assembled by a local artist to form, inside a square plastic bin, a living collage, Dr. Thomas writes that the ants "formed themselves into long, black, ropy patterns, extended like writhing limbs, hands, fingers, across the sand in crescents, crisscrosses, and long ellipses, from one station to another," and continues, "Thus deployed, they were watched with intensity by the crowds of winter-carapaced people who lined up in neat rows to gaze down at them. The ants were, together with the New Yorkers, an abstraction, a live mobile, an action painting, a piece of found art, a happening, a parody, depending on the light." He goes on, "I can imagine the people moving around the edges of the plastic barrier, touching shoulder to shoulder, sometimes touching hands, exchanging bits of information, nodding, smiling sometimes, prepared as New Yorkers always are to take flight at a moment's notice, their mitochondria fully stoked and steaming. They move in orderly lines around the box, crowding one another precisely, without injury, peering down, nodding, and then backing off to let new people in. Seen from a distance, clustered densely around the white plastic box containing the long serpentine lines of army ants, turning to each other and murmuring repetitively, they seem an absolute marvel. They might have dropped here from another planet." This reaction is imagined rather than real, because the ants actually died from the cold before he got to see them.

And then there is Thomas's outlook. He points out that no other large society in human history has achieved a standard of health as high as that which now exists in the United States. Life expectancy for Americans is now about seventy-two years, the longest that has ever been achieved in our kind of society. Most of us are in good health most of the time. Yet instead of enjoying this situation, we display a higher level of neurosis about our bodies and our health than ever before—we make ourselves sick by worrying about our health. It is estimated that 75 percent of the visits to doctors and clinics are by people who do not have anything organically wrong with them. They go to seek reassurance and psychological comfort. "It used to be quite different," Dr. Thomas noted recently. "When I was a medical student, it was not the habit of people, even in the more affluent parts of our society, to see the doctor regularly. One called the doctor when there was really something urgent going on. I remember being taught in a class by Dr. John Homans, an eminent surgeon at the

Peter Bent Brigham Hospital, in Boston, that if we ever encountered a patient from Down East Maine we should pay special attention, because these people *never* went to the doctor, and if one of them came into the office you had to take it for granted that you had a medical emergency on your hands." That was at a time when the health of everyone, including people from Down East, was immeasurably inferior to what it is now. Clearly, something has gone wrong. Something has become blurred in our view of our bodies and ourselves, and *The Lives of a Cell* is in part an attempt to illuminate this change and to restore things to their proper perspective. Here, for instance, is Dr. Thomas dealing with current attitudes toward personal hygiene:

Watching television, you'd think we lived at bay, in total jeopardy, surrounded on all sides by human-seeking germs, shielded against infection and death only by a chemical technology that enables us to keep killing them off. We are instructed to spray disinfectants everywhere, into the air of our bedrooms and kitchens and with special energy into bathrooms, since it is our very own germs that seem the worst kind. We explode clouds of aerosol, mixed for good luck with deodorants, into our noses, mouths, underarms, privileged crannies—even into the intimate insides of our telephones. We apply potent antibiotics to minor scratches and seal them with plastic. Plastic is the new protector; we wrap the already plastic tumblers of hotels in more plastic, and seal the toilet seats like state secrets after irradiating them with ultraviolet light. We live in a world where the microbes are always trying to get at us, to tear us cell from cell, and we only stay alive and whole through diligence and fear. . . .

In real life, however, even in our worst circumstances we have always been a relatively minor interest of the vast microbial world. Pathogenicity is not the rule. Indeed, it occurs so infrequently and involves such a relatively small number of species, considering the huge population of bacteria on the earth, that it has a freakish aspect. Disease usually results from inconclusive negotiations for symbiosis, an overstepping of the line by one side or the other, a biologic misinterpretation of borders.

Here is Dr. Thomas writing about the return of astronauts from the moon:

There is ambiguity, and some symbolism, in the elaborate ritual observed by each returning expedition of astronauts from the moon. They celebrate first of all the inviolability of the earth, and they reënact, each time, in stereotyped choreography, our long anxiety about the nature of life. They do not, as one might expect, fall to their knees and kiss the carrier deck; this would violate, intrude upon, contaminate the deck, the vessel, the sea around, the whole earth. Instead, they wear surgical masks. They walk briskly, arms up, untouching, into a sterile box. They wave enigmatically, gnotobiotically, to the President from behind glass panes, so as not to breath moondust on him. They are levitated to another sealed box in Houston, to wait out their days in quarantine, while inoculated animals and tissue cultures are squinted at for omens.

It is only after the long antiseptic ceremony has been completed that they are allowed out into the sun, for the ride up Broadway.

A visitor from another planet, or another century, would view the exercise as precisely lunatic behavior, but no one from outside would understand it. We must do things this way, these days. If there should be life on the moon, we must begin by fearing it. We must guard against it, lest we catch something. . . .

There are pieces of evidence that we have had it the wrong way round. Most of the associations between the living things we know about are essentially cooperative ones, symbiotic in one degree or another; when they have the look of adversaries, it is usually a standoff relation, with one party issuing signals, warnings, flagging the other off. It takes long intimacy, long and familiar interliving, before one kind of creature can cause illness in another. If there were to be life on the moon, it would have a lonely time waiting for acceptance to membership here. We do not have solitary beings. Every creature is, in some sense, connected to and dependent on the rest.

As it happens, the people in our society who seem to be the least infected by the health neuroses are doctors and their families. In *The Lives of a Cell,* Dr. Thomas speculates on why this is so. Since writing the book, he has received numerous letters from internists and general practitioners indicating that his hunch was right. He proposes a questionnaire that he would like to see answered by "the well-trained, experienced, middle-aged, married-with-family internists." He writes:

How many times in the last five years have the members of your family, including yourself, had any kind of laboratory test? How many complete physical examinations? X-rays? Electrocardiograms? How often, in a year's turning, have you prescribed antibiotics of any kind for yourself or your family? How many hospitalizations? How much surgery? How many consultations with a psychiatrist? How many formal visits to a doctor, any doctor, including yourself?

I will bet that if you got this kind of information, and added everything up, you would find a quite different set of figures from the ones now being projected in official circles for the population at large. I have tried it already, in an unscientific way, by asking around among my friends. My data, still soft but fairly consistent, reveal that none of my internist friends have had a routine physical examination since military service; very few have been x-rayed except by dentists; almost all have resisted surgery; laboratory tests for anyone in the family are extremely rare. They use a lot of aspirin, but they seem to write very few prescriptions and almost never treat family fever with antibiotics. This is not to say that they do not become ill; these families have the same incidence of chiefly respiratory and gastrointestinal illness as everyone else, the same number of anxieties and bizarre notions, and the same number—on balance, a small number—of frightening or devastating diseases.

In conclusion, Dr. Thomas writes, "The great secret, known to internists and learned early in marriage by internists' wives, but still hidden from the general public, is that most things get better by themselves. Most things, in fact, are better by morning."

From the time I read Dr. Thomas's book, I wanted to learn more about him—to find out what in his medical experience led to his outlook and philosophy. As one might imagine, Thomas is an extremely busy man, but, over a period of weeks, we had several long conversations in his office, which is on the top floor—the twentieth—of Sloan-Kettering. Dr. Thomas is a youthful-looking man in his mid-sixties, with a trace of gray in his brown hair. He is about six-feet tall and of squarish build, and is given to wearing three-piece suits of the type that, at least in my day, were referred to as "Ivy League." All in all, he resembles a Yankee ship captain. There is something about him that immediately inspires confidence, and while he has never had an ordinary medical practice, it is easy to picture him as a wise, no-nonsense family doctor. His conversation, like his book, is full of dry humor. His office is a large, bright place with long windows, some of which face the East River. There are various medical books and scientific journals neatly arranged on shelves. On one wall is a large painting—*The Sleeping Duck,* by Maurice Graves—and on another are a number of framed diplomas: for his Bachelor degree, received from Princeton in 1933; for the 1937 cum laude degree from Harvard Medical School; and for honorary degrees as well (Yale, '69, and the University of Rochester, '74). On a table in one corner stands an electric coffeemaker, and pipes of various sizes and shapes lie on his desk and on tables.

At our first meeting, I asked Dr. Thomas to tell me about himself, and he began at the beginning. "I was born in Flushing, New York, on November 25, 1913," he said. "On my father's side, we are of Welsh origin. His family came over here some generations ago—they are said to have been hod carriers. There were names in the family like Thomas Lewis Thomas, my grandfather's, and Lewis Smith Thomas, an uncle's—typical Welsh names that had been in the family for generations. My father, however, was named Joseph Simon Thomas. There had been a tradition of the Welsh language in the family, but by my father's time it had narrowed itself down to a single word"—at this point, Dr. Thomas emitted a sound that sounded something like "weegh"—"which he used to say at Christmas when he was carving the turkey. He never knew what the word meant, and I have never been able to find out. My mother's family were farmers from New England—she came from Connecticut—and, like many people from that part of the country, claimed some sort of Mayflower ancestry. As far as I know, it has never been proved."

"Was there a scientific tradition in the family?" I asked.

"My father was a doctor, but his father ran a mill—a flour mill, I think—in New Jersey," Dr. Thomas replied. "Other than my father, there were, as far as I know, no scientific people in the family. My father graduated from Princeton in 1899, and his medical training was at Columbia Physicians and Surgeons; he did his residency at Roosevelt Hospital. He moved to Flushing in 1905 or 1906 to set up his practice. Flushing was, he felt, a nice village to set up a practice in—a long carriage ride from Manhattan—and it was thought to be an area that might grow."

I asked Thomas if being the son of a doctor had given him the idea in his childhood of becoming a doctor.

"Yes, at that time, it did," he told me. "In those days, it was the normal thing in communities the size of Flushing for doctors to have their offices in their homes. The front part of our house was my father's office, and the waiting room, for patients. We lived upstairs. My earliest memories are of patients filling the waiting room; I also remember being taken—from the age of three or four on—in the Franklin sedan that he drove on house calls, and waiting for him while he visited patients. My impression was that he was having a good time in this work. It was obviously engrossing and exciting, and being a doctor looked to me like a great deal of fun."

This feeling must have been shared in the Thomas household. Thomas has three older sisters—Nancy, Ruth, and Edith—and a younger brother, Joseph, Jr. Ruth Thomas Feldmann is the widow of a doctor, and his brother is a practicing internist in upstate New York.

"Early in my father's career, he was a general practitioner, as I guess every doctor was at that time," Thomas continued. "Later, he trained himself as a surgeon while continuing his general practice, and in the mid-1920s he became a Fellow of the American College of Surgeons. He decided that he was going to do nothing but surgery, and announced to his patients that that was how things were going to be. That was the way one got to be a surgeon then. Later on—much later on—the requirements became formalized, with stipulations about the length of time that one was required to do this or that kind of surgery, at this or that kind of institution."

Turning the talk back to himself, Dr. Thomas said, "I went to grammar school in Flushing, at P.S. 20, and after a year of high school in Flushing I transferred to the McBurney Day School, in New York—a long subway ride from home. I graduated from high school and went to college when I was sixteen. I wasn't a prodigy of any kind, but I was a bright kid and skipped some grades. It seemed a good idea for me to get to college as soon as I could, because I knew I had a long pull after that—going on to medical school."

Although Thomas took the standard premedical course at Princeton, which included a good deal of chemistry, physics, and biology, his interest in being a doctor flagged during his first three years and revived only in his senior year, when he began to study advanced biology. At this time he began reading and writing poetry. Ezra Pound and T. S. Eliot were his favorites, and he wrote poems "in bursts"—first for the *Princeton Tiger,* and later, when he was an intern, for publications like the *Atlantic Monthly.* He still writes poetry, which over the years has been published in several magazines.

In 1933, when Thomas began his studies at Harvard Medical School, doctors were for the most part not taught to cure disease, and were not particularly expected to bring about cures. "The great thing was to find out the *name* of the disease and then to let nature take its course," he told me. "We were taught to make very accurate diagnoses, so that accurate prognosis was possible. Patients and their families could be told not only the name of the illness but also, with some reliability, how it was likely to turn out. We were taught how to give 'supportive treatment,' which consisted in large part of plain common sense: good nursing care, appropriate bed rest, a sensible diet, the avoidance of traditional nostrums and patent medicines, and a measured degree of trust that nature, in taking its course, would very often bring things to a satisfactory conclusion."

Dr. Thomas went on, "We were given a thin blue book called *Useful Drugs,* which had about a hundred pages. In my father's time, by the way, the textbooks on therapeutics were far more elaborate than anything I ever saw, and the pharmacopoeia he used was enormous—filled mostly with plant extracts that were to be combined following meticulously worked-out formulas. Three-quarters of a century ago, it was the general belief that a highly skilled physician was one who could write a long Latin prescription that combined eight or ten different drugs. None of this could conceivably have had any beneficial effect. In retrospect, most of it must have been totally worthless. This was beginning to be recognized by a few physicians at that time, notably William Osler. Dr. Oliver Wendell Holmes felt that if one took the whole pharmacopoeia that was then in use and threw it into the ocean the health of the country would be improved. By the time I became a medical student, all this had been replaced by a sort of therapeutic nihilism."

Dr. Thomas smiled. "Now the pharmacopoeia has got big again," he said. "The field is entirely different. A great many drugs have been developed since the 1940s about which we have some real, solid scientific information, and most of them—not all of them—are genuinely effective in the treatment of disease. I have a suspicion that the drugs that are most widely used—which are, of course, the tranquilizers, or so-called tranquilizers—are more like placebos than any of the other things that we use these days. I am not sure that for the ordinary patient, with a transient anxiety, these drugs have much, or any, effect beyond their placebo effects. If someone is told he is going to be given a drug that will make him feel tranquil, placid, and relaxed, he's almost bound to feel that way—for a while, anyhow."

After medical school, Dr. Thomas did a two-year internship on the Harvard Medical Service at the Boston City Hospital.

"Everybody's general health in the late 1930s was what we would consider these days to be very bad indeed," he told me. "People were subject to dangerous, often lethal infections that we have forgotten all about. In the City Hospital in the winter months, during my internship, every other bed was filled by someone with lobar pneumonia—pneumococcal pneumonia. The mortality rate varied with the type of the pneumococcus—between 20 and 50 percent. This was a very bad disease. Tuberculosis was common, and miliary tuberculosis and tuberculous meningitis were 100 percent fatal. We were still seeing typhoid fever. We had streptococcal septicemias and staphylococcal infections. They filled the wards, and people died of them all the time. Most of that run of disease has now vanished. One reason is that we have antibiotics that can prevent or terminate most of these infections.

"And yet people did not worry at that time in quite the same manner that they worry now. There was not that sort of fixed anxiety about health which one senses today. The health problems that now seem more conspicuous as threats are, in a way, more frightening to most people than the very acute short-lived but conclusively settled disease problems we used to have. Lobar pneumonia was a real threat to life, but either it cured itself or you were done for within ten days, and the same was true of most of the other major infections. Tuberculosis and syphilis were the two main long-drawn-out chronic diseases. Now, I think, people in general are more worried about other long-drawn-out diseases that are potentially fatal—especially cancer, but also heart disease, stroke, and rheumatoid arthritis. A lot of people are fearful about the kind of illness that is associated chiefly with old age. More of us have a real run at a long life span—more of us have the opportunity of going on into advanced age. That accounts in part for the anxiety.

"But there is something else going on that I do *not* understand. There is a tendency for people to become not just dissatisfied but positively apprehensive when small imperfections in the system turn up: things that are clearly self-limited, that are going to get better in a day or two, now cause more anxiety than they used to—respiratory infections, gastrointestinal upsets, a headache, various pains and aches that are part of normal living. These now tend to be interpreted as danger signals. All the patent medicines that television commercials spend so much time on are responsible, or partly responsible, for the population's growing much more psychoneurotic about health than was once the case. Our people seem to feel constantly threatened by physical disintegration and failure, and that is something that I simply do *not* understand. As you look around, you see that most of the people you know are in reasonably good shape. We are not foundering as a population. We have too much heart disease, and heaven knows we have too much cancer. But surely we have a great deal less in the way of what used to be lethal disease. We're *very* much better off than we once were."

Dr. Thomas paused, and then continued, "Onc useful thing that could be done, which we have never done well, is to have more of the real contemporary information about human physiology and, maybe, pathology somehow incorporated much earlier into everybody's education. I am always astonished by how little my most highly educated and intelligent nonmedical friends actually know about the functions of the human body. They have the most bizarre ideas of the way things really work. And much of it is not all that complicated. Quite a lot of good, hard factual data could be made a part of everybody's store of information—in high-school education, in college—and yet we don't do this well. We have students learning what we call hygiene and what are regarded as good habits of living, but they don't seem to know what the liver is all about or what the spleen is up to or how the lungs and the heart really function. We don't know anything as much about these things as we really would like to. The liver, for example, is still a considerable mystery. But we know an awful lot more than we did when I was a medical student forty years ago, and I would have thought that the level of general education would keep up with this, and that more of today's information would be known and understood by everybody—not reserved, as it is, just for the professionals."

Dr. Thomas reached for a pipe and continued, "I think what people in general fail to perceive about the human body is that most of the things that seem to go wrong get better by themselves. Most illness is self-limited. The vast majority of illnesses that affect young people, for example, are self-limited. Because of the masterstroke of the antibiotics, the threat of infectious disease has been largely eliminated, and what young people are left with are mild, transient episodes of illness—illness that goes away by itself.

"We now have an enormous bill for health in general in this country. When you add up the costs for doctors, hospitals, drugs, and so on, we spent over $120 billion on health in 1977, and it has been estimated that if we keep on the way we have been running for the last decade the country will be spending as much as $250 billion on it in the 1980s, and that's getting to be a very significant part of the gross national product. A lot of that is hospitalization cost, but a lot of it is also the cost of seeing the doctor, and the cost of the things that happen almost automatically after one has seen the doctor—the laboratory tests, the X-rays, and the coming back for a review and a final checkup. All that runs into money. I have a hunch that quite a lot of it is really not necessary. It has become a sort of social custom with us.

"I think it would be much easier to defend that part of medicine if we really did have a strong base of preventive medicine. If only it were true that by detecting illness earlier we could do something to prevent it or to correct it! Regrettably, that isn't true, except for a small number of diseases. There are a few varieties of cancer which if you see them early on you can do something effective to stop—once and for all. This is also true of glaucoma, perhaps of hypertension, but that's about it. The bulk of the serious illnesses that we still have with us run their natural course, and they are sufficiently mystifying to medicine so that we can't do much to stop them. If the doctors of this country had more time to see patients who are genuinely in need of their services and could spend less time seeing patients who do not, when they come to the doctor's office, really have much of anything the matter with them but are there simply to be checked up, I think we would have much less complaining about the doctor shortage. Far and away the majority of events that are reported to doctors as illnesses get better by themselves; they don't require anything special in the way of medical care. Unfortunately, they are frequently—far too frequently—treated with antibiotics, which are hardly ever necessary. They get better with bed rest. They get better in the morning."

In his two years at the Boston City Hospital, Dr. Thomas had become interested in specializing in diseases of the brain, and wanting to learn neurology, he became a resident-trainee at the Neurological Institute of Columbia Physicians and Surgeons, in New York, where he remained for the next two years—until 1941, which was when he concluded his formal clinical training. "By then, I knew that what I really wanted to do was research in medicine—that I wanted to stay in what we all called academic medicine," he said. "I went back to Boston, back to Harvard, and back to the City Hospital, where I had a fellowship to study meningitis infections of the brain." The fellowship paid $1,800 a year, and this affluence enabled Thomas to marry Beryl Dawson, a Vassar student, whom he had met at a dance when he was a sophomore in college. She lived in Kew Gardens and he lived in Flushing—Queens communities close to each other and to Manhattan, where they were married on January 1, 1941. The Thomases have three daughters: Eliza, a library assistant at the Widener Library at Harvard; Judith, who is a technician in a cell-biology laboratory at the Massachusetts Institute of Technology; and Abigail Thomas Luttinger, a poet, who is married to the theoretical physicist Joaquin Luttinger, now chairman of the Physics Department at Columbia University. Mrs. Thomas functions as an editor—Thomas calls her his collaborator—for everything that her husband writes for publication. A month after the Thomases were married, they went to Halifax, Nova Scotia, with a group of Harvard Medical School faculty members to study and help deal with an epidemic of meningitis that had broken out there. Mrs. Thomas worked in the bacteriology laboratory as a voluntary technician while her husband studied the effects of sulfadiazine, a new drug that turned out to be the most satisfactory drug for meningitis that had been found up to that time. That was one of the few times that Dr. Thomas has done clinical research involving patients. "As an experimental pathologist," he noted, "most of what I've done is to try to explore analogies for human disease in conditions in animals. I've pretty much stayed in the laboratory, with experimental animals."

After a month in Halifax, Dr. Thomas and his wife returned to Boston, and remained there until 1942; then they moved to New York, where he joined the Naval Medical Research attached to the Rockefeller Institute For Medical Research. The Rockefeller Institute had been established by John D. Rockefeller in 1901, with what was at that time a novel goal—to study the biological basis of disease. (Most of the other institutes devoted to medical research, such as the Pasteur Institute in Paris, had been set up to do for specific diseases.) Thc institute was then—and, renamed the Rockefeller University, is now—situated along the East River, on a tract of land that at the time of its purchase was being used to graze goats. Its organization was developed in 1902 by the American physician Simon Flexner, who suggested that the institute have an attached research hospital that could house sixty patients; the hospital was dedicated in 1910. Thomas M. Rivers, who was appointed its director in 1937, became a Navy captain during the war and was the head of the unit to which Dr. Thomas was assigned. The Rockefeller Hospital was equipped with almost every kind of medical specialist except a neurologist, so Dr. Thomas was taken on as a neurologist. "I was called to active duty in 1942, after I had had just over a year in Boston, and I came down here with the expectation of being a neurologist for the Navy," he told me, "but I ended up doing virology in Rivers's laboratory. Subsequently, in 1944, we went overseas, to Guam, and I had a virology lab of my own there, and then another virology lab that we set up in the field on Okinawa at the time of the invasion there."

I asked why the Navy, in the middle of the war, had set up a virology laboratory on a Pacific Island that had just been invaded.

"The Navy wanted to be sure that it would not get caught with unexpected epidemics among the men, and that the men didn't get into areas where there might be virus or rickettsial diseases that they hadn't counted on," Dr. Thomas replied. Howard Taylor Ricketts was an American pathologist who showed in 1906 that Rocky Mountain spotted fever was carried by ticks and who discovered the microorganism—called rickettsia, after him—that caused it. "Most of the group that had been assembled from the staff of the Rockefeller Institute was assigned to one or another of the possible areas of hazard that the Navy had in mind. My own assignment was scrub typhus, which was known to exist in the Pacific, although it was not known whether it existed on Okinawa. The laboratories on Guam served as base laboratories from which people went out to various areas where questions had come up. I was sent in to Okinawa on April 2, 1945, the day after the invasion, with a group of about a dozen scientists, which included virologists; rickettsiologists; parasitologists looking for the worms that cause schistosomiasis, which is due to a parasitic worm, carried by snails; some entomologists interested in mosquitoes and mites and various other kinds of biting insects; and even several mammalogists and ornithologists. I had the distinction of being the only person to go over the side of the ship and wade ashore with a box of a hundred white mice over my shoulder. We were going to use them for isolating the rickettsiae if we found any."

Dr. Thomas continued, "As it turned out, there wasn't any scrub typhus on Okinawa, and had never been any, but there was

Japanese B encephalitis, which was endemic on the island. The first cases broke out among the men and, simultaneously, among the natives sometime in June. We succeeded in isolating the virus in our field laboratory, and it was identified by the base laboratory in Guam. Albert Sabin, representing the Army, had come to the island at about the same time, and he and I had laboratories that collaborated on the problem during the six months I stayed there. Sabin had already, in the States, developed a vaccine against Japanese B, since it was expected that we would run into that kind of encephalitis at one place or another in the Pacific. Vaccination would have been feasible, but, as it happened, the outbreak simply smoldered along, involving not many people either in the native population or among the troops, so no action was necessary to cope with it. We got evidence that the virus was probably carried in some of the birds on Okinawa. We found antibodies in some of the birds living on the island, and we found that all the adult horses on Okinawa had very high levels of the specific antibody against Japanese B virus. In one experiment, we were able to demonstrate that the virus could be transmitted to young horses and could be recovered in the blood over the next several days. That suggested to us that perhaps mosquitoes or other insects were carrying the virus back and forth between horses and men, and perhaps between birds and other animals. Since that time, it has become clear that various species of birds do carry this virus, and it's known that horses also carry it. As it happened, encephalitis was the only major health problem that we ran into."

By the time Dr. Thomas left for the Pacific, he had already acquired a good deal of experience in dealing with virus-carrying birds. His first research problem at Rockefeller had dealt with isolating the psittacosis virus—the one that brings on parrot fever. "It is one of the most dangerous organisms to work with that I have ever touched," Dr. Thomas said. "It seems to me, these days, that a great deal more concern is being voiced about recombinant DNA, whose hazards, I think, are very largely imaginary, and too little attention is being paid to how successful work in the past has been on agents where it was necessary to do research and where the research was very hazardous and where, on balance, very few episodes either of infection of the laboratory workers or of infection of anybody else ever occurred. We tend to exaggerate the danger of spreading potentially hazardous agents." He and his colleagues found that a great many of New York's pigeons have high concentrations of antibodies against parrot fever—meaning that the birds are, or have been, infected by the disease. According to Dr. Thomas, mice, rats, and cats in the city also carry viruses related to but less virulent than psittacosis.

"During the time that I was there, I became involved in the studies going on at the Rockefeller on primary, atypical pneumonia, which was then called viral pneumonia," Dr. Thomas went on. "This was a major problem in the armed forces in the early 1940s. There was a considerable amount of apprehension about it, because it was a disease that spread rapidly through our troops here and abroad, and resulted in a lot of incapacitation and some deaths. The Navy was very much concerned about it, so a group of us at the institute, who were then in the Navy, went to work on the problem of viral pneumonia. One of our contributions was the isolation of an organism—not a virus but a streptococcus—which we named streptococcus MG. That's because the first patient we got it from was named McGinnis. We were interested in that organism because it was found that a majority of patients with viral pneumonia developed high concentrations of a specific antibody against that streptococcus within two weeks after the pneumonia infection. This suggested to us that perhaps the streptococcus itself was implicated in the causation of the disease. As it turned out, no such connection would be established, and we never did find out why it was that patients with that type of pneumonia selectively made an antibody against that particular streptococcus and not against any other. One side product of that investigation was that the laboratory test for antibodies against streptococcus MG had some usefulness as a diagnostic test for viral pneumonia."

In 1946, Dr. Thomas left the Navy and returned to a laboratory at the Rockefeller Institute. "I thought I would be staying there, but then I was offered a position in the Department of Pediatrics of Johns Hopkins, in Baltimore, with an opportunity to do research on rheumatic fever in children," he told me. "I was not a trained pediatrician—my training had been a combination of internal medicine and neurology—but I decided to go, because of the interest I had in that particular disease and in infectious diseases in general."

By the 1930s, it had been conclusively established that rheumatic fever is an aftereffect of a primary infection of a specific streptococcus—the so-called group A hemolytic streptococci. People do not get rheumatic fever unless they have been infected by the streptococcus, and this discovery has led to the use of antibiotics that have all but eliminated the disease in this country. "There is something bizarre about this illness," Dr. Thomas noted. When rheumatic fever develops, it does not do so for ten days or two weeks after the first infection, usually in the throat—by which time the original throat infection has already disappeared. "The disease has some similarities to a generalized allergic reaction," Thomas said. "It is as though the streptococcus infection had turned on some kind of highly inappropriate immunological reaction, in which the host's own defense mechanisms have brought about damage to his organs—in this case, the heart. I became interested in the use of cortisone, which was then becoming available, in the treatment of rheumatic fever. I was one of the early skeptics about the effectiveness of cortisone for this. We did some clinical research on it. Cortisone appeared to switch off the symptoms of rheumatic fever—patients felt and looked a great deal better—but the amount of heart-valve damage was about the same. The patients were not fundamentally better off than patients who had not been treated with cortisone.

About that time, I encountered another phenomenon involving cortisone that turned out to be quite interesting. My colleagues and I discovered that rabbits treated with cortisone became extremely vulnerable to infection by a variety of different bacteria, particularly the streptococcus. It was later shown that human patients became vulnerable to infections after cortisone treatment, and cortisone was recognized as being potentially a very hazardous form of therapy. I spent some of the next few years trying to find out why cortisone had the remarkable effect of at the same time damping down inflammatory reactions—suppressing inflammation—in general and opening up approaches on the part of bacteria for becoming established and producing overwhelming and often lethal infections. That problem remains unsolved. We're still not certain how cortisone does eliminate host defenses against infection. This continues to be one of the most interesting problems in immunology." It was Dr. Thomas's work on the experimental pathology of rheumatic fever and related pathological phenomena which led him to suspect that much, if not most, of the damage caused by disease results not from anything intrinsically toxic about the invader but from a misreading of signals by the host, which then unleashes an overwhelming and inappropriate defense reaction; these defense mechanisms may cause the damage.

After two years at Johns Hopkins, Dr. Thomas was asked by Tulane University, in New Orleans, to set up and run a Division of Infectious Disease at Charity Hospital there. He spent two years there, too, and was then offered a chair that spanned the Departments of Internal Medicine and Pediatrics at the University of Minnesota, in Minneapolis, where he was again able to study rheumatic fever in children. It was at Minnesota that Dr. Thomas did his work on cortisone. He remained at Minnesota until 1954, when he was asked to come to New York University to set up and become the chairman of a Department of Pathology. He also became a member of the New York City Board of Health and spent some time at City Hall, trying to persuade the city to rebuild Bellevue Hospital, which eventually came about 1958. "It sounds as if I could never quite hold a job," Dr. Thomas remarked. "I was head of the Department of Pathology until 1958, when I was appointed chairman of the Department of Medicine and of the NYU Divisions in Bellevue Hospital. I stayed in that job for a record time—until 1966, when I was made dean of the NYU Medical School. In 1969, when I joined the Yale University School of Medicine's Department of Pathology, I had an opportunity to stop doing administrative work for a while, but after three years I was appointed dean of the Yale Medical School, and I remained in that job until I came here to Sloan-Kettering. In retrospect, I must say that I enjoyed the kind of administrative work I did, which mainly involved putting something together—a division or a department that had to be either started from scratch or put back together again. I am not sure that I would enjoy it now, in the kind of decade we are in. You have to remember that during those years—beginning in the 1950s—the National Institutes of Health had just gone into large-scale operation, and it was possible to build new enterprises with a certain amount of security as far as being able to fund research was concerned, which is no longer the case."

After a pause, Dr. Thomas went on, "I've always kept a laboratory going, wherever I was. I did so when I was dean at NYU and when I was dean at Yale, and I do so now. I would miss not having a laboratory. It's more than just a hobby—it doesn't feel right to me not to have something going on in research." Indeed, our conversations were occasionally interrupted by telephone calls from Miss Dorothy McGregor, who was collaborating with Dr. Thomas on his research, and who gave him up-to-the-minute reports on what was happening downstairs in the lab. "I get to our laboratory for part of just about every day," Thomas told me. "It's not a big enterprise. It's a very small laboratory, with no one in it but Miss McGregor and me."

I asked what he was working on at present, and he replied, with evident enthusiasm, "We have three problems. The first one concerns the morphology of bacterial protoplasts and their role in infectious disease. The second involves endotoxins, which I have been working on since 1950, when I was at Minnesota. The third has to do with olfaction." Protoplasts are bacteria that have lost their walls and acquire different properties because of the loss, he explained, and he went on, "We are studying the possibility that protoplasts may be implicated in certain chronic infections in man. It is something I have been interested in for the past eight years." Endotoxins are bacterial parasites whose poison is released only when the endotoxin is broken up. This release appears to happen by a sort of allergic reaction in the host cells. It is as if the host cells misread the innocent nature of the endotoxins, and the defense reaction of the cells releases the very poisons that are most dangerous to the cell. There is a breakdown in communication between the host and the parasite. Dr. Thomas told me that this was the sort of thing he had in mind when he wrote that disease "usually results from inconclusive negotiations for symbiosis, an overstepping of the line by one side or the other, a biologic misinterpretation of borders." He went on to say, "What we have been doing is to study the reactions that occur in the cells of the horseshoe crab, *Limulus,* when they come in contact with endotoxins. This has turned out to be a useful way of detecting very small amounts of endotoxin. Anything we can learn about endotoxins is particularly relevant to what we are doing here at Sloan-Kettering, since it has been known for a long time that endotoxins have a rather special necrotizing effect on rapidly growing tumors."

Dr. Thomas paused, and then continued, "As for olfaction, I made a prediction several years ago in a talk I gave, and then in a paper I published, that there had to be an odorant that we and other animals emit which marks each of us biologically as our individual selves. There is a lot of evidence for this in animals other than man. From behavioral studies, it is known that bullheads are able to recognize the smell of other bullheads, as individuals—by name, as it were. Minnows can recognize other minnows as individuals—not just as minnows as a class but as persons. And there has always been evidence that tracking hounds can distinguish one human being from another by scent. There are two papers, by the way, that indicate that identical twins cannot be clearly distinguished by tracking hounds. My prediction was that in the best of worlds it would turn out that the genetic basis for that kind of marking would be very closely related, if not identical, to the genetic determinants that mark our cells and tissues—to the immunologic machinery that is responsible for the specific rejection of skin grafts and grafts of other tissues between individuals. It seemed to me that nature would not be as parsimonious as nature ordinarily is if two totally related genetic systems had been set up for this kind of self-identification.

"Here at Sloan-Kettering, Dr. Edward Boyse and his wife, Jeanette, breed special congenic inbred strains of mice. Mice so inbred that they differ from each other by only *a single group of genes:* the so-called H-2 locus, which is responsible for histo-compatibility—the immunological marking. Jeanette Boyse noticed from observing their breeding behavior that these mice were able to distinguish between themselves and other mice differing only by H-2. Male mice were seen to show a preference for mating with females of a particular H-2 type, showing that mice can sense each other's H-2 histo-compatibility types. The conclusion that we finally arrived at was that the gene locus is in fact the same for the compatibility of tissues and for the olfactants that identify oneself as oneself."

Dr. Thomas added, "And Dr. Peter Andrewes, of our group, is now involved with the head of the tracking-hound service, or whatever it's called, in the Baltimore Police Department—Sergeant Tom Knott. The hounds are used mostly to smell out explosives or narcotics, and they are extremely sensitive. We are trying to find out if they can tell the difference between closely related individuals, to find out if the equivalent gene locus in human beings—the H-LA locus, it's called—is also responsible for our self-marking odorants. We are also trying to find out if we can train the hounds to tell the difference between two congenic lines of mice just by sniffing them or their cages or something connected with them. I don't know how *that* is going to come out. All this bears on what we are trying to do at Sloan-Kettering. It has a bearing on the problem of the genetic determinants of self-marking mechanisms, which is one aspect of the cancer problem. Any information that bears on the human immunologic system which recognizes foreign cells or foreign tissues is relevant to what we are up to here. We manage to keep those three things going. They have their ups and downs, but if we are lucky we hit one going up when the others are going down."

The collection of essays that led to *The Lives of a Cell* began almost accidentally. In 1970, when Dr. Thomas was at Yale working on experimental pathology, he was invited to give a lecture for a symposium at Brook Lodge, outside Kalamazoo, Michigan. This symposium was one of a series of informal seminars for people in the biomedical field, bringing together about forty participants from all over the world. They are run by the Upjohn Company. This particular one was on the general subject of inflammation. "I had not been working closely in the field myself for several years," Dr. Thomas told me during one of our talks. "So I decided to put together a rather general paper about inflammation and what I thought it might signify as a process in biology. The talk was printed up by the sponsors of the meeting, and Franz Ingelfinger got a copy of it."

Franz Josef Ingelfinger was, until 1977, the editor of the *New England Journal of Medicine,* a weekly publication of the Massachusetts Medical Society. It is one of the oldest and most distinguished American medical magazines. "Ingelfinger is an old friend of mine," Dr. Thomas explained. "We were interns and house officers together at the Boston City Hospital back in 1938. He had been in academic medicine himself—had been a professor at the Boston University School of Medicine before he became editor of the *Journal,* about ten years ago. When he got a copy of my talk, he wrote to me and said that he liked the way some of the things in it had come out, and would I be interested in doing a column for the *New England Journal of Medicine.* He spelled out the terms. I would have to stay within the limits of two pages—around 2,000 words. In return for that, he promised that whatever I sent in, if it was accepted, would not be edited by anybody. And that was irresistible, so I decided I would try it.

"At the time this happened, we were at Yale, but we had a house near the Marine Biological Laboratories at Woods Hole, Massachusetts, where I also had a small lab. My wife and I used to drive up to Woods Hole on Friday nights from New Haven. I would start writing Friday night and finish an essay Sunday night. I wrote four or five essays, and then I wrote Ingelfinger a letter and said that maybe that was enough—I ought to stop it. But he wrote back and said no, I'd better not, they were kind of interesting, and would I keep on. After I got started, I found that I couldn't write any differently. I had to write something that was around 2,000 words—not much more than that. When I found that I was writing something longer than that, I also realized that it wasn't going well. That length seemed to be very congenial to me. I was never quite sure in advance what I would be writing about, and I usually became convinced during the month that I would not have an essay that month. But as it turned out, I was really pretty good about it. I got about thirty of them done before the book came out. I was surprised when I found that someone thought that the essays could be a book, because I hadn't realized that they were that much connected to each other."

I remarked to Dr. Thomas that a number of the ideas in *The Lives of a Cell* had been entirely novel to me. One of the things that had struck me most was that the book offered a completely different way of looking at the nature of disease. I was one of those people Dr. Thomas wrote of who believed that there was something inevitable about disease—that it was somehow an essential part of human existence. I found it perfectly reasonable that the astronauts returning from the moon should be quarantined until it was known whether or not they had brought back some unknown infectious agent. Consequently, I was especially impressed by the observation in *The Lives of a Cell* that "it takes long intimacy, long and familiar interliving, before one kind of creature can cause illness in another." I asked Dr. Thomas about this and about the nature of disease in general.

"I have a notion that to be a pathogen, and especially to be a pathogen for us, really takes a long period of interliving and a lot of familiarity," he replied. "We don't really know very much about most infectious diseases beyond knowing the names of the infectious agents. We don't know as much as we should know about how they cause tissue destruction—how they cause what we recognize as disease. But it does look as though part of the mechanism were in the elaboration by invading organisms of signals, and of misleading signals. There's a certain molecular mimicry that seems to go on in disease. I will give you several examples of what I have in mind.

"The diphtheria toxin is a chemical so closely related to a component of our own cells involved in protein synthesis that it's mistaken for that by the cell and locks onto a part of the apparatus for making proteins and shuts the process off. But in order to make the toxin the diphtheria bacillus itself has to have been infected by a virus. A perfectly healthy diphtheria bacillus is innocuous. It's not until it has been invaded by a phage—a bacterial parasite—and has incorporated the genetic content of the phage into its own genetic content that it has the information for making diphtheria toxin. There has been speculation that the information that comes in by way of the phage may itself have come long ago from eukaryotic cells, which are nucleated cells like ours. To a degree, therefore, it looks as though some infectious diseases were really accidents—biological accidents.

"Take meningitis—meningococcic meningitis, which is epidemic meningitis. When you first look at it, it looks as though the meningococcus were a predator bent on getting into the meninges and invading the central nervous system of human beings. But when you analyze epidemics of this disease, it turns out that an actual case of meningitis is a relatively rare event. In a population undergoing an epidemic, virtually everyone is infected by the meningococcus. Most of the people have positive throat cultures—virulent meningococci living in the mucosa at the back of the nose or the throat—and many of these people have in their blood detectable antibodies against the organism. But only a very few of them—a few thousand out of hundreds of thousands—get meningitis. For most of them, it's a mild, really unapparent respiratory-tract infection. In my view, most infectious disease doesn't make any biological sense. It's a sort of no-win game. The meningococcus has nothing to gain that I can discern from getting into the bloodstream and infecting the central nervous system. The organism gets along fine by living in the nasal and throat mucosa of the host. It must be even more of a catastrophe for a meningococcus to catch a man than it is for a man to catch a meningococcus."

Dr. Thomas paused for a few moments. "Then, there are infections caused by the Gram-negative bacteria—'Gram-negative' being a taxonomic term based on the Gram stain," he went on, and he explained that the stain process is named after Hans Christian Joachim Gram, a Danish physician who died in 1938. "All such organisms stain red," he said. "Most of the Gram-negative organisms possess at their surfaces endotoxins that are incorporated into the surface wall and are released when the organism dies. Nobody really knows how endotoxin works, but what it does when it is injected into animals, or when we are exposed to it through an infection, is to cause a complex series of reactions, any one of which would be classed as an important part of illness-high fever, and when the dose is large enough shock develops, with the bottom dropping out of the blood pressure, followed by coma and death. There are peculiar kinds of hemorrhagic reactions that occur in the skin and in a lot of internal organs, and it looks as though a lot of important physiological mechanisms had been unhinged by this toxin. But when you look at endotoxin closely it's difficult to find anything about it that is really toxic, in the sense that it inflicts a primary kind of damage. It doesn't poison the cells if you put it in cell cultures. It looks more like a signal than like a toxin. Indeed, Dr. Robert Good and I found out long ago that if you pretreat animals with cortisone and then inject them with an endotoxin, the signal is no longer read and none of these things occur in the animals. Endotoxin is innocuous. I think the way it works is that cortisone has the effect of shutting off—for a time, anyway—the inflammatory reaction, and defense reactions in general. I think it will turn out to be the case that the reaction to an endotoxin is the mobilization of a whole lot of different, unrelated defense mechanisms, which are all turned on at once—something like a panic. It's just a great big biological mistake. But it is also a kind of prototype for infectious disease in general. There are not many infectious diseases in which you can prove actual direct damage to cells or tissues by the invading organism."

Dr. Thomas gave me a specific example of how the system's defenses work against the organism they theoretically protect. "There is a virus disease that is called lymphocytic choriomeningitis," he said. "It has been studied a lot in mice, because it is endemic in mice. The full-blown disease, which can be produced by injecting a mouse with the virus, is a very serious acute meningitis, with a lot of lymphocytes, or lymph cells, suddenly appearing in the spinal fluid and covering, as though invading, the surface of the brain and the inside of the ventricles of the brain. It's a lethal infection that proceeds very rapidly, and the animals all die. If you give the virus to a newborn mouse, however, or, better yet, to a mouse that is still a fetus, then none of these things occurs. The virus lives in the animal while he is growing up. It replicates; you can always demonstrate a living virus in the bloodstream, but there's no meningitis and no brain disease. Later on, there are some ill effects. The animals age more rapidly, and they're more likely to develop kidney disease, but they don't get this special disease that is so characteristic when an animal is confronted with the virus for the first time after he is past the neonatal period. It was shown by several people some time ago that if you take one of these animals that have the virus in them but are not reacting to it, and transfer to him lymphocytes or a lymph node from a normal animal, he will get the classic disease. This is because these newly transferred lymphocytes *are* able to recognize the virus and react to it. The animal that is infected in the newborn period has become what we call tolerant, and the cells of his immunological system are unable to recognize the virus as a foreign agent.

"The tolerant mouse that has been given a lymph node from a normal mouse dies because of the invasion of his central nervous system by lymphocytes—not by the virus. The virus itself is not doing him any harm, as far as one can tell; the virus doesn't damage brain cells—it doesn't damage the surface of the brain. His own immunologic reaction to the virus is what kills the animal. The tolerant animals, which are able to sustain a heavy burden of the live replicating virus, eventually die with kidney disease. But even this is due to the immunologic reaction. That is so because they succeed in making so much circulating antibody to the virus that over time it forms with the virus, and the insoluble complexes are deposited in the glomeruli—the blood vessels—of the kidney, and the kidney is eventually destroyed by that. This does the animal in, but he's really being done in by his own defense mechanism."

This view of infectious disease as the misreading of signals between the invader and the host has some fascinating implications. For the signals to be misread, they must first be read. If a signal is too strange, it will simply be ignored. This appears to explain why certain diseases are specific to certain forms of life, and not to others.

"It has been a preoccupation of infectious-disease investigators for a great many years to try to set up animal models for specific human diseases, but, as it has turned out, that is very difficult to do," Dr. Thomas told me when I asked about this. "For instance, no one has succeeded in reproducing classical lobar pneumonia in experimental animals. One can infect them with the pneumococcus and produce septicemia, which in the case of mice actually kills them, but one cannot reproduce the human lung disease—the lobar pneumonia. Pathogens are highly specialized for a particular host. As an example, I have worked with, and still work with, peculiar organisms that are called mycoplasmas. A mycoplasma is structurally somewhat like a small bacterium, but it doesn't have any cell wall. It is a separate species of creature—unrelated, really, to bacteria. It is the smallest free-living form of life; that is, it is bigger than a virus, but it is the smallest thing that can get along on its own—unlike a virus, which has to live inside a cell. These organisms are fascinating precisely because of the high degree of host specificity that they demonstrate. There is one mycoplasma—one strain of the gallisepticum mycoplasma—that causes serious illness in turkeys. The turkey, I can assure you, is one of the hardest animals to work with I have ever had in a laboratory. If you infect a turkey with this strain of gallisepticum mycoplasma, it develops a systemic disease of the small arteries, with lesions that closely resemble a human disease called polyarteritis nodosa. They look something like the lesions that also occur in malignant hypertension. Every artery in the turkey brain is involved in this disease. It is a highly lethal infection, and only a very small dose of mycoplasma is required to produce the full-blown disease and death. When this disease occurs, I am told, the turkey flocks are got rid of, not because it is hazardous to human beings, but because it is an economic hazard to turkey raisers. One cannot produce any disease, or any lesion at all, in a chicken, or in a mouse or a rat or a guinea pig, or in any other animal that I have tested, no matter how much of the mycoplasma you give it. This organism is adapted to the turkey. It will infect only the turkey. In fact, the injected mycoplasma in the turkey take up residence only in the walls of the arteries, and this implies that there is some kind of recognition system going on there. There is something that labels the mycoplasma and the arterial wall as matching in some way. The mycoplasma migrates selectively to this one particular part of the turkey, and to no part of any other species of bird or mammal. I think that this strain of gallisepticum mycoplasma, which is the only one that can produce arterial lesions—the others are a common cause of upper-respiratory disease in turkeys—may be a mutant. In my opinion, one population of gallisepticum mycoplasma underwent a mutation somewhere in California and developed this new property of being able to infect, and to elaborate a toxin within, the arterial wall. Perhaps a bacteriophage got into that particular mycoplasma and has conferred on it this property of producing a unique disease."

"All this being the case," I asked, "what are the chances that a virus or something like a virus from the moon or Mars—if such a virus exists—would cause pathogenicity in earthly life forms?"

"I think that if there should be a live agent on the moon or on Mars it would in all probability not be equipped with membrane markers of its own that could be recognized by us, or by any other animals, or by plants," Dr. Thomas told me. "I doubt whether we would have anything nutritionally to offer that would be special for such a microbe. It might live on the earth, all right, because it might be able to make use of things that are in the air or in the soil, but I doubt very much whether it could learn its way, all that quickly, into our kind of complexity, and form the kind of attachment that has to be formed for disease to occur. I can't imagine its being able to catch hold of our kind of life and get into it with the intimacy that is required for what we know as pathogenesis—to get things sorted out, to recognize the parts of a cell with the exactness that microbes do display when they are infectious agents. I just cannot imagine this taking place."

Dr. Thomas continued, "In fact, I think the general rule in nature which most microorganisms follow when they are living in association with what we call higher forms is for the arrangement to be symbiotic. In the first place, most bacteria, of which there are, I suppose, millions of tons on the planet, are in no sense pathogens for anything. The occupation of most microbial life is to recycle things—get rid of dead matter by causing it to decay to the point where it can go back into the cycle of life. But there is a small minority of that enormous population of microbes that live as symbionts with eukaryotic forms, and they're terribly important.

"The classic example, which everyone cites for this, is the rhizobial bacteria that live in the root nodules of leguminous plants and are responsible for nitrogen fixation. There is a lot of specificity to this—certain species of rhizobial bacteria live in certain species of plants, and can live only in those species. If you extract lectins from different varieties of legumes, you find that the lectins are attached preferentially to the cell surfaces of the rhizobial bacteria from a particular plant, and are not attached to the cell surfaces of rhizobial bacteria from unrelated plants. There's a very carefully worked out molecular recognition system between the microorganism and the host, and it's entirely beneficial to both parties. And there are organs in insects that are made up of bacteria. Nobody knows what they do, except that they're indispensable. If you get rid of them with antibiotics, the insects die off after a while. Cockroaches have highly specialized bacteria living in their tissues all the time. Obviously, ruminants—cows and the like—couldn't survive without the bacterial populations that they have for breaking down cellulose. Termites couldn't live without having protozoans that are responsible for digesting their wood, and those protozoans couldn't do that without the bacteria that they contain as symbionts—the bacteria that provide the enzymes to do the job. That is, I think, a much more common arrangement in nature than disease—than infectious disease."

After a pause, Dr. Thomas went on, "I think that every now and then organisms that may have been set up originally to live in that kind of accommodation have been subject to a mistake.

Either the organism itself has overstepped the line between symbiosis and parasitism or something has happened to the host mechanism which allows this funny kind of alarm reaction to occur in the face of signals coming from bacterial populations. There really aren't many diseases in which the bacteria come in and absolutely overwhelm the host and destroy it cell by cell. There are some, though. Anthrax is a disease in which the host is killed by the sheer mass of the invaders. The anthrax bacillus grows so rapidly that it almost occludes the circulation, and also it elaborates some kind of toxic material. But for most infectious diseases it's not that way at all. It's the overreactivity on the part of the host that does the damage."

This discussion raised the question of how many diseases—serious, incapacitating diseases—there are that lack adequate remedies. Is the list finite, or does a new disease occur to replace any that has been eradicated?

"I would put the list at about twenty-five," Dr. Thomas said. "There are a lot more than twenty-five illnesses, but I am narrowing it down to what I would regard as the major problems—the ones that are chiefly responsible for incapacitation or premature death. That list has from twenty to twenty-five items, which include cancer, stroke, heart diseases, schizophrenia, arthritis, kidney failure, cirrhosis, and the degenerative diseases associated with aging. There are a lot of diseases that are uncommon—in our kind of society, at least—and so are not major problems for us. In the so-called underdeveloped world, there are diseases that turn out to be, when they are different from ours, infections that could probably be controlled by our kind of technology, if it were available to the people. There are also in these societies protozoal infections and parasites and nutritional diseases that we don't have and don't understand. But, even so, the list of diseases is finite, contrary to what most people think. People seem to think that there is just no end to the things that can go wrong with us."

I next raised the question of maladies like Legionnaires' disease and some of the new lethal diseases in Africa that seem to arise out of nowhere.

"It is true, and I guess it has always been true, that from time to time what seem to be totally new kinds of disease appear," Dr. Thomas observed. "But I don't think anybody can ever be certain that a disease is really new, because of the possibility that it has always been around but we haven't had the diagnostic techniques available to uncover it or that it has been so infrequent that it has gone unnoticed. It's been a long time, though, since we have had anything brand-new turning up in the way of disease—something unfamiliar that affected a whole lot of people and that hadn't been seen before. I think that the last time that happened was in 1917 and 1918. In those years, there was the epidemic of Economo's disease—a worldwide epidemic of encephalitis that resulted in parkinsonism in many people—and there was also the 1918 epidemic of influenza. Both of those were apparently new diseases. When we look back, we see that there have always been outbreaks of new diseases, or diseases not recognized before, which turned up and swept across the scene and then vanished, like the sweating illness of medieval times. It is sometimes said, because every now and then a new disease is reported, like Legionnaires' disease, that we are always at risk of being endangered by a new disease—that even if we do enough good science and develop a technology that is as effective for, say, stroke or heart disease or arthritis as the antibiotics are for things like pneumonia and meningitis, there will be a new disease to come in and take its place. But there is no evidence for that. We really have got rid of a fair number of very important diseases, or nearly got rid of them, in the last half century. As far as I know, nothing has come in to take the place of diphtheria or pertussis or poliomyelitis or lobar pneumonia, which is now an uncommon disease, or epidemic meningitis. We seem to be getting along fine without them. I think there is something primitive about the idea that it is our destiny to have a kind of allotment, a quota of disease—that we are always obliged to have a certain number of diseases, and that if we get rid of today's diseases there will be new ones ready to take over. I think that's probably nonsense."

Dr. Thomas went on, "Some of the diseases we have almost got rid of are diseases about which there would have been general agreement less than a hundred years ago that they were the major scourges of mankind. The best example of this, I think, is tuberculosis, which really endangered everybody in the nineteenth century and in the first part of this century—even people who were able to live in what for that time was affluence. There were some things that could be done about tuberculosis when it was recognized that it was a bacterial infection. That occurred in the 1890s. It then became possible to identify early cases and isolate them; sanatoria were set up and the incidence of the disease began to fall. But it wasn't until streptomycin, and the improved antibiotics that were developed after streptomycin, that it became possible effectively to get rid of the disease. Tuberculosis still occurs, but now when it does we can turn it off. We can turn it off even when it is in what used to be its most lethal form. The most lethal disease that I ever had to deal with when I was in training was tubercular meningitis. It was one of the few diseases that everyone knew then was 100 percent fatal—everyone knew there was nothing at all to do about them. And now that disease can be turned off, in just no time at all, with proper treatment. But a century ago we were using the same kind of rhetoric in talking about tuberculosis that is now used in talking about rheumatoid arthritis and the rest. We thought that tuberculosis was an environmental disease. We had theories about air, bad air, moist air, night air being responsible for it. Sunlight was supposed to be good for it. All kinds of diets were cooked up to influence it one way or another. Bed rest. Go to Arizona. Go to Switzerland. But the fact of the matter was that once one had a technology that would get rid of the tubercle bacillus, that was that.

"And since we have got rid of a few very important diseases, I think there is good reason to believe that we can keep at it. I don't think we've reached a point now—at least, not for any reason *I* can guess at—where we're suddenly stymied, and stuck with the diseases left. I have to confess, however, that a lot of people in my field do think that we're now stopped. It has become something of a popular notion to say that the diseases we are left with now that we have got rid of the major infections are in some sense so complicated and so multifactorial, as the term goes—that they have something to do with the environment, or have something to do with stress and the pace of modern living—that we can't do anything about them until society itself is remade. I can't see any evidence for this. I simply can't take that point of view very seriously—at least, not as long as we are as ignorant about the mechanisms of those diseases as we are. We really don't know anything at a deep level about the mechanisms of heart disease, or cancer, or stroke, or rheumatoid arthritis. We can make up stories about them, and it could be, I suppose, that they do have multiple causes, and are due to things we can't control in the environment. If that's true—if that should turn out to be true—that would be quite a piece of news. Because it has never happened before. Every disease that we do know about, and for which we have really settled the issue, so that we can either turn it off, switch it off, or prevent it once and for all—every such disease turns out to be a disease in which there is one central mechanism. There may be a lot of other things going on, and maybe a lot of things that we don't know about have to do with predisposition to the disease, and maybe a lot of things aggravate the disease once it is established, but there is always a chairman of the committee. In the case of pneumonia, it's the pneumococcus, and in the case of tuberculosis it is the tubercle bacillus, and in pellagra it's a single vitamin deficiency. And I have a hunch—of course, I can't prove it—that it will turn out to be that way for cancer and probably for coronary occlusion, probably for stroke and probably for the kind of kidney disease that develops into chronic renal failure. Although there may be a lot of things going on, there will be one central, master mechanism for each of them, which we may be able to change when we learn what it is. I think we are a long way from this. We've got a lot more to learn, God knows. Except for the infectious diseases, our understanding of disease mechanisms is still primitive. But I think that the diseases that are now the important ones—that everyone is concerned about—are perfectly respectable biological problems for research."

It is all but incredible how far we have come, especially in the last century. Dr. Thomas called my attention to an article that Edward H. Clarke, a professor of medicine at Harvard University, wrote in 1876, on the occasion of the American Centennial. After summarizing the progress of medicine over the previous fifty years, Professor Clarke concluded that its major scientific accomplishments up to that time were studies showing that some patients with typhoid fever and typhus had actually gotten better by themselves—that there was actually such a thing as self-limiting disease. Previously, it had been the general view that unless a doctor did something—anything—a patient would die. "In response to that, doctors down the centuries *did* do things," Dr. Thomas said. "They did a lot of spectacular things, like bleeding and cupping and purging. Any procedure that anybody concocted in his mind could be tried out on human beings—the most frivolous and irresponsible kind of human experimentation, based on nothing but trial and error and usually resulting in precisely that sequence." By 1900, the idea that doctors should or could cure disease had been all but abandoned, and had been replaced by the therapeutic nihilism that Dr. Thomas refers to. This lasted until the 1940s, when it was replaced by an entirely new view of medicine.

"Up to the time the sulfonamides and penicillin came on the scene, there was much more acceptance of the fact of disease, and the risk of dying prematurely," Dr. Thomas remarked. "We had got used to the situation, after all, because we had always had it.

It was not unusual when I was a child for there to be a death or two among the younger members of the families of friends around town. Old cemeteries are full of tombstones of young children. People died of tuberculosis and pneumonia and meningitis. This was taken more or less as a fact of life, a risk that everybody ran. Nobody expected it ever to change. And then, suddenly, it did change, and it became clear that you really could do something about lethal disease. It became a natural thing for people to feel that if we could get that far in so short a time maybe we could go the whole distance—could do something, right away, about cancer and heart disease and stroke and all the rest. Our expectations rose very considerably beyond the capacity of medicine for meeting those expectations."

But what if we can eventually meet those expectations? What if the mechanisms that underlie cancer, heart disease, and the rest can be uncovered, and simple technologies like immunization can be found for dealing with them? Would this make us happy? Certainly it would ease the burdens of our later years. We might, as Dr. Thomas is fond of saying, "age dry, and then wear out all at once, like Oliver Wendell Holmes's one-hoss shay." But is fear of illness really at the root of our cultural sadness? After all, 75 percent of the people who visit the doctor in our country do not do so because they are organically ill. Many people visit the doctor because they are fundamentally unhappy. What does this say about us and about the times we live in?

Dr. Thomas has thought a great deal about such questions. On my last visit with him, he gave me something to read that he had written for an address to the American Association for the Advancement of Science in 1977. It said:

> These are not the best times for the human mind. All sorts of things seem to be turning out wrong, and the century seems to be slipping through our fingers here at the end, with almost all promises unfilled. I cannot begin to guess at all the causes of our cultural sadness, not even the most important ones, but I can think of one thing that is wrong with us and eats away at us: we do not know enough about ourselves. We are

ignorant about how we work, about where we fit in, and most of all about the enormous, imponderable system of life in which we are embedded as working parts. We do not really understand nature, at all. Not to downgrade us; we have come a long way indeed, just to have learned enough to become conscious of our ignorance. It is not so bad a thing to be totally ignorant; the hard thing is to be part way toward real knowledge, far enough to be aware of being ignorant. It is embarrassing and depressing and it is one of our troubles today.

It is a new experience for all of us, on unfamiliar ground. Just think, two centuries ago we could explain everything about everything, out of pure reason, and now most of that elaborate and harmonious structure has come apart before our eyes. We are *dumb*.

This is in a certain sense a health problem after all. For as long as we are bewildered by the mystery of ourselves and confused by the strangeness of our uncomfortable connection to all the rest of life, and dumbfounded by the mystery of our own minds, we cannot be said to be healthy animals in today's world.

We need to know more, and there it is. To come to realize this is what this seemingly inconclusive century has been all about. We have discovered how to ask important questions, and now we really do need, as an urgent matter, for the sake of our society and its culture, to obtain some answers. We now know that we cannot do this any longer by searching our minds, for there is not enough there to search, nor can we find the truth by guessing at it or by making up stories for ourselves.

Thomas concludes, "T. H. White wrote something about this, in *The Once and Future King,* where Merlin was advising the despondent young Arthur: 'The best thing for being sad,' he said, 'is to learn something. Learn why the world wags and what wags it. That is the only thing the mind can never exhaust, never alienate, never be tortured by, never fear or distrust, and never dream of regretting. Learning,' he said, 'is the thing for you.' "

PART III

Fact and Fantasy

ONE of the laws of Arthur C. Clarke states that the activities of a really technologically advanced society would appear to us like magic. It has occurred to me, and to others before me, in the same vein—that a really novel scientific idea always looks crazy. Of course, it can simply be wrong-crazy. The interesting situation is when the idea is crazy, but right. The three pieces of writing collected here are concerned with ideas that are thought to be, or have been thought to be, crazy—but have proven to be or appear to be right. The first is a profile of Arthur C. Clarke, who has had enough crazy—but right—ideas to fill several heads besides his own, the most celebrated one being the communications satellite, which he thought of at a time when many of his elders were arguing that orbital space flight was impossible, or, at best, hopeless. The second piece is concerned with the theoretical possibility that automata can be designed so that they reproduce themselves. This idea is for some people repellent, for others absurd, and for still others a fascinating prospect. I am of all three minds, and the piece explains why. The third is my fictional attempt to describe both Gödel's theorem and the state of mind it induces in me when I attempt to understand it. I remember my feeling of disbelief when I was first told that there must exist mathematical statements which are both true and unprovable—crazy—but right; that, crudely speaking, is what Gödel's theorem says. These bits of writing deal with the outer limits of the scientific experience, the wilder shores of the human imagination—crazy, but right.

6

EXTRAPOLATORS: ARTHUR C. CLARKE

According to Muslim legend, Adam, after he was forced to leave the Garden of Eden, made his way south through India and onto the island of Ceylon (Sri Lanka), where, later joined by Eve, he settled down to propagate the human race. Another version of the legend has it that Adam, for his disobedience, was hurled from Heaven and, as penance, stood for a thousand years on one foot on top of Adam's Peak, a remarkable 7,360-foot mountain that rises in a precipitous rock pyramid from the jungles and tea estates of central Ceylon. Indeed, at the summit of the mountain there is a shrine built around a rock that bears a curious footprint-like indentation, which Muslims take to be the footprint of Adam, and which Buddhists—who constitute about 70 percent of the present population of the island—take to be the footprint of Gautama Buddha. Adam's Peak, which is visible for miles out to sea, was the landmark that guided the first Europeans, the Portuguese, to the island in 1505. They were looking for cinnamon, and they found, in addition, the decaying fragments of a great civilization created by the Sinhalese—the "lion race," of Indian descent, who had flourished on Ceylon for nearly 2,000 years. The Sinhalese had constructed fantastic cities like Anuradhapura, which is said to have had, by the first century A.D., a population of several million, but by the time the Portuguese arrived the ancient cities had been largely reclaimed by the jungle, and the kingdom of the Sinhalese was weak and divided. The government of Ceylon had retreated to Kandy, in the highlands near Adam's Peak. (Kandy served as Mountbatten's headquarters during the Second World War and was one of the three principal locales of the film *The Bridge on the River Kwai,* which was made entirely in Ceylon.) The Portuguese were never able to subjugate the island, although they claimed it in the name of their king. But their influence still manifests itself through the presence of the Catholic Church and in the many Portuguese family names, like de Silva. It was the Portuguese who named Colombo (after Christopher Columbus). However, their tenure was short-lived; beginning in the early seventeenth century, the Dutch East India Company, in alliance with the Kandy kings, drove the Portuguese out. At the end of the eighteenth century, the Dutch, in turn, gave way to the British East India Company; the British captured Colombo in 1795, and in 1802 Ceylon became a Crown Colony—a status it retained until 1948, when it became fully independent.

Ceylon, which is shaped like a teardrop and hangs just off the southeastern coast of India, is about half the size of Great Britain, and what traces of European influence are left in Ceylon are very largely British. The country's second language is English. (Its first language is Sinhalese, and its third is Tamil.) It was the British who built the modern port of Colombo. (Colombo is not a natural harbor, and the British literally created one.) And it was the British who introduced the cultivation of tea and rubber—currently the two principal exports—to Ceylon. The island's population is now about 12 million, of whom some 7,000 are Europeans, but most of these are British—mainly retired planters who fell in love with the jungles, the magnificent seacoast, and the highland country, and could not face going back to the rigors of the English climate after their tea and rubber estates were taken over by the Ceylonese. However, the most singular of the British expatriates now living in Ceylon is not a planter at all but one of the truly prophetic figures of the space age, Arthur C. Clarke.

Clarke, a remarkably youthful-looking sixty (when he is wearing his glasses, which he needs because of myopia, he presents the vaguely mischievous appearance of a benign and stupendously energetic blue-eyed owl), has the rugged physique and constitution of a farm boy, which he once was. He has been well known as a science-fiction writer and a scientific prophet for over a quarter of a century. (In 1959, he made a bet that the first man to land on the moon would do so by June, 1969.) However, it is only in the last few years—especially since he and Stanley Kubrick wrote *2001: A Space Odyssey*—that he has become widely known to the general public. He became even more widely known during the first flight to the moon, when he served as one of the commentators assisting Walter Cronkite in his coverage of the event for the Columbia Broadcasting System. Cronkite had been a Clarke fan for many years, and Clarke has done a number of television broadcasts with him, beginning as far back as 1953. In following the Apollo 11 flight, Clarke made some dozen appearances. During an early one, Cronkite asked him if he would mind explaining the ending of *2001,* and Clarke answered that he didn't think there was enough time—then or later. He went to Cape Kennedy with the CBS team, and at the moment of the launch, as he told a friend on his return, he, like everyone around him, burst into tears. "I hadn't cried for twenty years," he said. "Right afterward, I happened to run into Eric Sevareid, and he was crying, too." After the launch, Clarke returned with the rest of the CBS crew to New York and spent most of the next several days in and out of the CBS studios, watching the flight and, from time to time, going on camera. The actual landing on the moon was, in many ways, the fulfillment of a life's dreaming and prophesying. "For me, it was as if time had stopped," he said later.

Clarke's oeuvre appears to have a special appeal for young people, and he spends a good deal of his time, when he is in the United States, lecturing to university audiences, who seem to have been familiar since childhood with his science-fiction classics, such as *The City and the Stars, A Fall of Moondust, The Sands of Mars,* and *Childhood's End,* as well as with his superb popular-science books, such as *The Promise of Space* and *Profiles of the Future.* Clarke once received a letter from a sixteen-year-old amateur filmmaker that began, "You are the best sci-fi writer no doubt in the world (or at least in the North American Continent);" and went on, "Could you please do me a great favor (which I'll repay you when I become famous), and that is to think of an idea for a suspenseful sci-fi short movie (about ½ hr.), and let me have your permission to use it as a theme."

Anyone who talks with Clarke for very long discovers several characteristics of his conversation that make it both delightful and sometimes difficult to follow. He is afflicted with what he refers to as a "butterfly mind." His conversation flits in and out of subjects, passing from one to the other with the speed of an agitated lepidopteron. He will sometimes provide a transitional "exactly, exactly" while at the same time conveying the impression that he has not been paying much attention to any interjections the listener has been making in the vain hope of slowing him down. When he writes, however, his style is a singular amalgam of scientific erudition, speculative imagination, and a profoundly poetic feeling for the strange and only partly understood objects—stars, moons, planets, asteroids—that populate our universe. Many—indeed, most—science-fiction writers do not have any special scientific training. Clarke (like Isaac Asimov, who is a biochemist) is a notable exception. He served as an electronics engineer with the RAF (in 1945, while still in the service, he originated the idea of a communications satellite); he has a bachelor's degree (with First Class Honors) in physics and mathematics, and has done advanced study in astronomy; and for a spell he worked as an abstracter for a major scientific journal. There is little doubt that Clarke could, if he felt like it, earn his living by doing science. One of the reasons he doesn't is that it might get in the way of speculation, which, while consistent with generally accepted scientific principles, often pushes them a step or two farther than most scientists would be willing to go. In this connection, Clarke once remarked, "Since I don't have any scientific reputation to lose, I can say what I please without giving a damn about what the professionals think of it." However, the really distinguishing feature of Clarke's style is the sense of sadness and loneliness that man must feel over living for so brief a time in such a vast universe of which he can have so limited a glimpse. Kubrick once remarked, "Arthur somehow manages to capture the hopeless but admirable human desire to know things that can really never be known."

Since Clarke is himself an unswerving optimist, with an all but Buddhist reverence for life, the worlds that he creates, however strange, are usually basically benign. It is very rare in a Clarke story to find an "alien"—i.e., an extraterrestrial being—or an animal, or even a plant, that one feels one could not get on with, given a little practice. Moreover, the inanimate objects also acquire personalities. Here is Clarke describing the prehistoric earth in "The Sentinel," the short story that was transformed into *2001:* "Such was our own Earth, the smoke of the great volcanoes still staining the skies, when that first ship of the peoples of the dawn came sliding in from the abyss beyond Pluto. It passed the frozen outer worlds, knowing that life could play no part in their destinies. It came to rest among the inner planets, warming themselves around the fire of the sun and waiting for their stories to begin." And here is Clarke describing Alvin, the hero of *The City and the Stars,* in his first encounter with a strange robot: "None of the conventional control thoughts produced any effect. The machine remained contemptuously inactive. That suggested two possibilities. It was either too unintelligent to understand him or it was very intelligent indeed, with its own powers of choice and volition. In that case, he must treat it as an equal. Even then he might underestimate it, but it would bear him no resentment, for conceit was not a vice from which robots often suffered." Stanley Kubrick once summarized Clarke's gift by saying, "He can take an inanimate object like a star or a world, or even a galaxy, and somehow make it into a very poignant thing that almost seems alive."

Clarke has made his home in Ceylon more or less permanently since 1956. What drew him there, and keeps him there, is the sea. Ceylon is surrounded by some of the most magnificent coral reefs in the world—reefs that abound in tropical marine life. Since 1951, Clarke—despite the fact that he has never learned to swim properly—has been an ardent deep-sea diver and photographer in such diverse places as the English Channel; Clearwater, Florida; the Great Barrier Reef of Australia; and along the coast of Ceylon. Of the more than 500 articles, books, and short stories that he has written to date (Clarke has catalogued his output in several neatly handwritten looseleaf notebooks, with entries indicating when the work was done, who bought it, and how much was paid), about a quarter have to do with the sea. The rest include technical electronics articles (most notably his classic paper on the communications satellite, which appeared in the October, 1945, issue of *Wireless World*, and in which he proposed extraterrestrial relay stations, plotted the orbits required, and set forth in great detail the fundamental principles of the satellite's design); other scientific papers dating back to his RAF days; short stories like "The Sentinel" and "Before Eden"; novels like *Earthlight* and *Against the Fall of Night;* popular scientific books like *The Exploration of Space* and *Interplanetary Flight;* and, of course, the enigmatic screenplay for *2001*. Despite all this activity, which has taken Clarke to most of the centers of space technology in the Western world as a consultant, a lecturer, or an observer, the spell of the sea has always drawn him back to

Ceylon. He has summarized his feeling about Ceylon in *The Treasure of the Great Reef,* a book on deep-sea diving. In it, he describes the discovery and salvage of a seventeenth-century treasure ship that was sunk on the Great Basses Reef, off the south coast of Ceylon—an enterprise he undertook with Mike Wilson, a former British paratrooper, frogman, film producer, and professional diver, who introduced Clarke to diving and was his partner in underwater explorations. Of Ceylon, Clarke writes, "Though I never left England until I was thirty-three years old (or travelled more than a score of miles from my birthplace until I was twenty), it is Ceylon, not England, that now seems home. I do not pretend to account for this, or for the fact that no other place is now wholly real to me. Though London, Washington, New York, Los Angeles are exciting, amusing, invigorating, and hold all the things that interest my mind, they are no longer quite convincing. Their images are blurred around the edges; like a mirage, they will not stand up to detailed inspection. When I am in the Strand, or 42nd Street, or NASA Headquarters, or the Beverly Hills Hotel . . . my surroundings are liable to give a sudden tremor, and I see through the insubstantial fabric to the reality beneath." He concludes, "And always it is the same: the slender palm trees leaning over the white sand, the warm sun sparkling on the waves as they break on the inshore reef, the outrigger fishing boats drawn up high on the beach. This alone is real; the rest is but a dream from which I shall presently awake."

Clarke was born on December 16, 1917, in the small seaside town of Minehead, in Somerset, on the Bristol Channel. (His speech still carries the lilting accent characteristic of Somerset; "moon," for example, becomes "moo-un," with a melodic emphasis on the "moo.") Clarke's father was then a soldier at the front, and, to make ends meet, his mother, grandmother, and an aunt were running a boardinghouse, which still stands. After the war, Clarke's father invested in a farm nearby, and when Clarke was about five the family moved from the boardinghouse to the farm.

The first years there were a financial disaster, and this was compounded by the death of Mr. Clarke when Arthur was still in his early teens. By this time, there were three other children in the family—two brothers, Fred and Michael, and a sister, Mary. (Fred was until very recently a heating engineer, and is now the full-time director of the Rocket Publishing Company, a family enterprise that invests Clarke's European royalties in various underwater and film operations. Michael has settled on the family farm and is running it, and Mary, a former Royal Navy Nursing officer, is now married and at home.) Mrs. Clarke had to take over the farm herself to make a living for her four children. When I visited Clarke in Colombo, I met Mrs. Clarke, who was making her first trip to Ceylon. She turned out to be a delightful ruddy, gray-haired woman somewhere in her seventies (Clarke says she simply will not tell her children how old she is, and none of them has been able to find out), who shares with the other Clarkes I have encountered a general lucidity of mind and fluidity of speech.

Even a casual reader of Clarke's fiction cannot fail to be struck by the fact that its animals, robots, and aliens often appear to be more human than the human beings. Squeak, the first Martian discovered by the Earthlings in *The Sands of Mars,* and the lion that forms a charming friendship with a young boy in *The Lion of Comarre*—and even the squids in one of Clarke's stories, "The Shining Ones," to say nothing of the wonderful aliens in *Childhood's End* and the benign and melancholic ultimate intelligence, Vanamonde, in *The City and the Stars*—all have the appealing characteristics of domestic animals on a bizarre cosmic farm. Clarke is well aware of this and attributes it, in part, to the fact that he has always been more interested in things and ideas than in people. He also feels that it is attributable in part to his early life on the farm. In addition to the usual farm animals, the Clarkes were surrounded by dogs, which his mother raised professionally. Clarke recalls that at one point they had fourteen puppies in the house, which surged back and forth in waves between the rooms. "We were drowning in a sea of them," he remembers. Clarke keeps several dogs—large German shepherds—in his house in Colombo, as well as a number of wild birds, which lodge in the interior nests that the Ceylonese put in their dining and living rooms to bring good luck. Clarke will not kill any animal that is not actively venomous, and during my visit with him I aided in gently depositing out-of-doors a considerable assortment of tropical spiders of impressive size and girth, which, Clarke insisted, wouldn't hurt anything, and whose presence was necessary on the front lawn to preserve the delicate ecological balance. Indeed, Clarke's first interest in science came through animals—prehistoric animals. He recalls driving in a pony trap with his father, and his father's giving him a cigarette picture card showing a prehistoric reptile. Clarke was immediately fascinated, and began collecting the other cards in the series. Then, when he was about twelve, after a brief spell of fossil collecting, he suddenly discovered astronomy, and, as he puts it, "that was it." The next several years became a feast of reading astronomy books, copying down astronomical tables, and constructing telescopes. Clarke is a firm believer in the Freudian dictum that adult happiness lies in the fulfillment of unfulfilled childhood aspirations. As a child, he could not afford any scientific instrument that he did not make with his own hands. By way of compensation, he has now purchased a first-rate Questar telescope, which he sets up on his lawn in the evenings in order to watch the moon and planets and any satellites that happen to pass over Colombo. In addition, he owns a very advanced Zeiss microscope. (Not long ago he was given, by the Indian government, a setup for receiving television transmissions from satellites, and became the only person in Sri Lanka with television.) He and I spent a few afternoons peering at various objects through the Zeiss—a butterfly wing, a plant stem, and a slide of a cross-section of tissue from a human brain, which had been lent to Clarke by a local surgeon. Clarke told me that while he was writing the novel *2001* he used to devote a good deal of time to contemplating the brain tissue, and that this is what had inspired him to write of astronaut David Bowman's final transformation: "He seemed to be floating in free space, while around him stretched, in all directions, an infinite geometrical grid of dark lines or threads, along which moved tiny nodes of light—some slowly, some at dazzling speed. Once he had peered through a microscope at a cross-section of a human brain, and in its network of nerve fibers had glimpsed the same labyrinthine complexity. But that had been dead and static, whereas this transcended life itself. He knew—or believed he knew—that he was watching the operation of some gigantic mind, contemplating the universe of which he was so tiny a part." The one childhood ambition that Clarke has never realized is to own the most elaborate possible Meccano set—the British equivalent of an Erector set. But the ambition is still with him. "In my declining years, when I am too feeble to totter out to the telescope, I could build almost anything I wanted with a set like that," he says.

It was during Clarke's high-school days that he encountered the idea of space travel. He came across a secondhand book called *The Conquest of Space,* by David Lasser, and persuaded his aunt to buy it for him. (Some years ago, Clarke met Lasser in the United States and had the pleasure of thanking him for starting him off on the subject.) At about the same time—in 1935—he joined the British Interplanetary Society, a group of scientific amateurs who had gotten together in Liverpool two years earlier to study space travel. (Admitting a teen-age boy to membership was not as remarkable then as it would be today, for, as Clarke observed recently, "no respectable scientist would touch the subject with a barge pole." However, many of the early members of the BIS have gone on to become key figures in the British rocket program and in British postwar technology in general.) Clarke had always been outstanding at mathematics in school, and when he graduated—there being no money for a college education—he took the civil-service examination for a post as a government auditor. He passed (twenty-sixth among

1,500 candidates), was given a job with the Board of Education in London, and hated it. For some time, he had been writing science-fiction stories, and then, in 1937, he began to publish them in a science-fiction journal called *Novae Terrae,* which he and a number of like-minded amateurs printed on a mimeograph machine. "Almost all British science-fiction writers of that period got started the same way," he told me. "Home-operated mimeographs, home-grown articles and stories." In the somewhat carefree atmosphere that appears to have prevailed in the prewar British civil service, Clarke found that he was able to get a good deal of writing done during his working day. But then the war cut short his career as an auditor. Deciding that he wanted to remain as close to astronomy as possible, he joined the RAF. (Actually, he did not have to enlist, since as a civil servant he was in a reserved occupation.) Largely because of his poor eyesight, he was not allowed to fly; instead, he was sent into a new and top-secret project called RDF, which was concerned with what later became known as radar.

Clarke's going into radar was in many ways a turning point in his life. In the first place, it was his introductory encounter with the actual instruments of advanced technology; in the second place, it was his introductory encounter with professional scientists, as opposed to science-fiction enthusiasts and amateur rocket buffs. He was sent to Yatesbury, a bleak place on the moors near Stonehenge, where he went through a broad course in electronics and then a specialized course in radar itself. After taking this course, he was made an instructor—a job that he found immensely pleasant, since he enjoys teaching. Clarke spent a year or so at No. 9 Radio School, the radar school, and, having a good deal of spare time, he taught himself advanced mathematics and electronics. Out of this experience, he wrote his first technical paper, on Fourier analysis as applied to television wave forms, and it was published in *Electronic Engineering.* After working for a while on early-warning radar systems—so crucial in the air war over England—he was put to work on the installation of the first

Ground Controlled Approach landing system, a radar device developed by Luis Alvarez, who is now at the University of California at Berkeley and is a recent Nobel Prize winner in physics. Alvarez showed how radar beams, which had been used to locate planes, could also be employed to guide them down a glide path to safe landings in any weather. It is a basic system still in use at many airports. Clarke has always believed strongly that no part of a writer's experience should go to waste, so he used this experience to write *Glide Path*, his only non-science-fiction novel, and certainly the only novel ever written about radar. It is dedicated to Alvarez.

By 1945, Clarke was a flight lieutenant stationed a few miles from Stratford-on-Avon and teaching airmen to maintain the new GCA system. He was also writing and selling science-fiction stories in his spare time. It was during this period that he got his idea for a communications satellite. He outlined his conception of what has become one of the most important—and perhaps the most commercially profitable—inventions of the space age in a paper called "Extraterrestrial Relays," for which he was paid forty dollars by the British technical journal *Wireless World*. Clarke has summed up this episode in an essay called, somewhat plaintively, "A Short Pre-History of Comsats; or, How I Lost a Billion Dollars in My Spare Time."

Clarke has been in the business of scientific and technological prophecy for over forty years now, and from this experience he has evolved a set of laws and principles. There are three basic Clarke laws. (He once remarked that if three laws were enough for Newton they were enough for him.) The first Clarke law states, "If an elderly but distinguished scientist says that something is possible he is almost certainly right, but if he says that it is impossible he is very probably wrong." Clarke has confirmed this law by counting up the elderly but distinguished prewar astronomers who "proved," by portentous calculations, that space flight was technologically impossible. The second Clarke law was originally a simple sentence in his book *Profiles of the Future* but was promoted to a law by the translator of the French edition. It states, "The only way to find the limits of the possible is to go beyond them into the impossible." The third, and most recently formulated, Clarke law, which he made use of in writing the enigmatic ending of *2001,* states, "Any sufficiently advanced technology is indistinguishable from magic." In addition to the laws, there are several empirical principles, one of which Clarke feels is fully applicable to his 1945 *Wireless World* article on the communications satellite; namely, that in making scientific prophecies the tendency is to be optimistic in the short range and pessimistic in the long. At the time that Clarke wrote his *Wireless World* article, the V-2s had already fallen on London, so it was well-known that high-altitude rockets were a practical possibility. Clarke felt that they would be used as high-altitude research probes, and in 1944 he predicted that this would take place within a decade, which was somewhat optimistic. However, the communications satellite, he felt, would not come into existence for half a century or more, which was pessimistic, since Syncom 3, the first synchronous TV satellite, was launched on August 19, 1964. In his "Pre-History," Clarke has an interesting aside concerning that launching. He writes:

> This event, incidentally, is a good example of the perils that beset a prophet. In October, 1961, while moderating a panel discussion at the American Rocket Society . . . I had mentioned that the 1964 Olympics would be a good target to shoot for with a synchronous satellite. (I cannot claim credit for the idea, which I'd picked up in general discussions a few days earlier.) Dr. William Pickering, director of the Jet Propulsion Laboratory, was in the front row of my audience, and he was so tickled with the suggestion that *he* passed it on to Vice-President Johnson, speaker at the society's banquet the next evening. The Vice-President, in turn, thought it was such a good idea that he departed from his prepared speech to include it; so when "Profiles of the Future" was published in 1962, I felt confident enough to predict that most large cities would carry live transmissions from Tokyo in 1964. What I had failed to foresee was that, despite heroic efforts by the White House, the Communications Satellite Corporation, NASA, and the Hughes Aircraft Company (builders of Syncom 3), a large part of the United States

did *not* see the superb live transmissions from the Olympics, which were made available by this triumph of technology. Why? Because they arrived at an awkward time, and the networks did not want to upset their existing program and advertising arrangements!

In 1963, Clarke received the Stuart Ballantine Medal of the Franklin Institute for his conception of the communications satellite. He keeps a picture of the medal, which is awarded for advances in communications, on the wall of his study in Colombo. The original, a rather valuable chunk of gold, is stored in his bank.

Clarke's wartime experience convinced him that, whatever else, he was not going back into civil service. For one thing, he was just beginning to sell science-fiction stories to American magazines—in particular, to *Astounding Stories*. For another, he wanted to go to college and follow up his scientific training. He managed to get a financial grant and, early in 1946, entered King's College, London, from which he took, with First Class Honors, a B.Sc. degree in pure and applied mathematics and physics. (In at least one case, the educational process was reciprocal. Clarke introduced one of his professors, George C. McVittie, now of the University of Illinois and a well-known cosmologist, to the theory of rocketry.) While in college, Clarke completed his first published novel, *The Prelude to Space,* which he wrote during a summer vacation. His other extracurricular activities included the revitalization of the British Interplanetary Society, which had more or less dwindled away during the war. In 1946, he became its chairman, and soon thereafter managed to recruit one of its more remarkable members—George Bernard Shaw. This came about soon after Shaw read a paper of Clarke's entitled "The Challenge of the Spaceship," which Clarke had sent him just after—as it happened—Shaw's neighbor, Geoffrey de Havilland, had been killed during a test flight in which he was trying to fly faster than the speed of sound for the first time. On January 25, 1947, on one of his famous pink postcards, Shaw wrote to Clarke, "Many thanks for the very interesting lecture to the BIS. How does one become a member, or at least subscribe to the Journal? When de Havilland perished here the other day it seemed clear to me that he must have reached the speed at which the air resistance balanced the engine power and brought him to a standstill. Then he accelerated, and found out what happens when an irresistible force encounters an immovable obstacle. Nobody has as yet dealt with this obvious limit to aeronautic speed as far as I have read. G.B.S." Fortunately, Shaw was a better playwright than a physicist, and in his reply Clarke pointed out, as tactfully as possible, some of the Shavian misconceptions of the dynamics of flight, adding that, in any case, his lecture had been concerned with space flight, which takes place above the atmosphere. Shaw responded with a detailed two-page letter and then became a member of the BIS, remaining one for the rest of his life.

Clarke took his undergraduate degree in two years—his RAF training having given him a considerable head start—and graduated in 1948. Having done very well as a student, he was offered an additional year in which to study advanced mathematics and astronomy. Within a few months, he had become thoroughly disenchanted with the interminable minutiae of the latter. "Luckily, before I perished of boredom, the dean of the college wrote to me to say there was a job opening for me," Clarke said recently. This turned out to be the assistant-editorship of *Science Abstracts,* a journal put out by the Institution of Electrical Engineers, which performed the useful function of abstracting technical papers on physics from every scientific journal in the world. By 1948, there was a tremendous accumulation of papers, based on wartime work, which was just being published in the open literature. "I tore into this huge mountain of arrears," Clarke told me. "We had about a hundred abstracters, some of them multilinguists who would tackle anything, in any language, on any subject, whether they knew anything about it or not. We were rather at their mercy." Clarke did a good deal of abstracting himself, and even managed to get astronautics into the journal as a legitimate scientific subject. One of the most difficult jobs was indexing new subjects in physics, because, as often as not, as soon as these appeared they would also disappear, because some underlying experiment or theory was wrong. "Sometimes my objectivity as an abstracter was severely strained," Clarke recently recalled. "I can remember a paper from an Indian physics journal proving that no rocket could ever travel faster than the velocity of sound in the metal of which it was composed—complete nonsense, of course, which would put an end to any thought of space travel. I wonder what I indexed it under. And it would be interesting to know what that particular scientist is doing now. In any case, the job was ideal training for a budding science-fiction writer." During the nearly two years that Clarke remained on the staff of the journal, he was writing in his spare time, and when his income as a writer got to be larger than his income as an editor he decided to quit and become a full-time free lance. His first nonfiction book, *Interplanetary Flight,* was followed, in 1951, by *The Exploration of Space,* which was a selection of the Book-of-the-Month Club.

It was at about this time that Clarke discovered skin diving. Just after the war, British science-fiction writers and editors used to gather in a London pub off Fleet Street called the White Horse. (The White Horse has become the White Hart in Clarke's collection of fantasy stories called *Tales from the "White Hart.")* The meetings came to be regular Thursday-night affairs, and on one rainy Thursday night in 1950 Mike Wilson, who was then in his late teens—too young to be in a pub—wandered in by accident, looking for shelter from the weather. Wilson had been at sea in the Merchant Marine and was at the time working as a wine waiter in a London hotel. He was a keen science-fiction fan, and it was soon evident that he had come to the right place. He had done some skin diving in the Orient, and he told Clarke about it. Clarke was immediately seized by the idea, and he was soon taking skin-diving lessons in various London swimming pools, and diving with a rented aqualung in the English Channel. Shortly thereafter, on the bounty from the Book-of-the-Month Club, Clarke made his first trip to New York, and, at Abercrombie & Fitch, bought his first aqualung, which he put to use in the Weeki Wachee Springs, near Clearwater, Florida. (While in Florida, he also got married to a local girl after a courtship of a few days. The marriage was not a very happy one and ended in divorce some years later.) By 1954, Clarke's skin diving had become his major interest. "It was a hobby that got out of hand," he has said. In support of his obsession, he arranged with Harper & Row to do a book on the Great Barrier Reef, the world's largest coral formation, which lies off the northeast coast of Australia. Mike Wilson was sent ahead to reconnoiter the terrain. (He filled in his time by working as a professional pearl diver, out of Darwin.) Clarke followed on the S.S. *Himalaya,* an elderly P. & O. liner, which sailed from London through the Suez Canal to Aden, and then made a stop in Colombo. Clarke had an afternoon in Colombo and—again by one of those coincidences that seem to have characterized his life (he once remarked that "nothing bad can ever happen to a writer")—met the leading diver and underwater naturalist on the island, Rodney Jonklaas, a descendant of one of the original Dutch colonial settlers. It was Jonklaas who told Clarke about the potentialities for underwater exploration around Ceylon, and, as Clarke has said, "I decided that if I survived the Great Barrier Reef I would come back and do a book on Ceylon." Just before Christmas of 1954, the *Himalaya* arrived in Sydney, and Clarke spent the next year on the Great Barrier Reef, diving and photographing underwater life. This resulted in his first book on the sea *The Coast of Coral.*

Wilson had also stopped off in Ceylon on *his* way to Australia, and, like Clarke, had been attracted to the place. So, in 1956, they decided to move there. At first, they lived in a small place in a Colombo suburb near the water—just the two of them and a cook named Carolis. ("Carolis" is presumably a Latinized version of "Charles." Such Latinized names among the Ceylonese are a legacy from the Portuguese occupation.) In time, Wilson married one of the most beautiful Ceylonese girls on the island, Elizabeth Perera, and they—and Clarke—moved into a house in the fashionable quarter of Colombo which was owned by Elizabeth's parents. At the time of my visit, Clarke still rented the Perera house and Carolis was still part of the Clarke retinue. Though dour of countenance, he seemed a kindly man, the father of somewhere between nine and thirteen children—Clarke has lost count—and an expert in making Ceylonese curry, which is likely to incinerate the palate of any unsuspecting Westerner. In addition to Carolis, Clarke maintained a staff of two houseboys, a secretary with an impenetrable Tamil name that has been shortened to Thambi, and a Ceylonese boatman named Martin, who had been with Clarke and Wilson since the beginning, and who spends his time taking care of a garageful of aqualungs, flippers, air compressors, spears, masks, snorkels, and other such gear. Two more current residents were Hector Ekanayake, a Sinhalese diver and the former welterweight boxing champion of Ceylon, and Hector's fiancée. Clarke does not speak Sinhalese, so the house reverberated with a constant buzz of multilingual conversations that were all but unintelligible over the din of aqualung compressors and the general hammering that goes with the maintenance of the diving equipment. He has since moved into a large new house, and I have lost count of the number and variety of its various occupants.

Clarke's only listing in the Colombo telephone directory was under "Clarke-Wilson Associates," the outfit that he and Wilson founded to engage in underwater exploration. Although Clarke and Wilson are still close friends, Clarke-Wilson Associates is not functioning as such. Indeed, at the time of my visit, Wilson had entered what Clarke referred to as "his guru, or prophet, phase." He had taken to retreating to the hills near Kandy and meditating. The first time I met him at Clarke's, he was dressed in a sort of white prophet's outfit, and his close-cropped hair and long, thin face gave him a Svengali appearance. He was wearing a golden necklace with a rectangular slab attached to it, which Clarke later identified as a promotional ornament that M-G-M was then offering to the hippie crowd in connection with *2001*.

In the late 1950s and the early '60s, Clarke's life settled down to a pleasant routine of underwater exploration around Ceylon, writing (he produced, among other things, the novel *The Deep Range,* about the farming and ranching of the plants and animals in the sea in the future), and making three-month lecture tours in the West. In March, 1961, a discovery that Wilson made while diving off the Great Basses Reef changed their lives for the next few years. Wilson and two young boys had gone to the reef to make a film about the boys' discovery of diving and underwater swimming. Clarke, who had remained in Colombo, was surprised to see the three of them come back furtively carrying a tin trunk. Wilson, carefully closing the door to Clarke's office, opened the trunk, revealing two perfectly preserved miniature brass cannons, along with a heavy, stonelike lump that, on careful inspection, turned out to be a solid mass of silver coins, fused together by long immersion in seawater. The three swimmers had discovered a sunken treasure ship. Most of the coins were in prime condition and bore a Muslim date whose Christian equivalent was 1702. So far, Clarke and Wilson have not been able to establish the exact circumstances of their wreck's final voyage—or, indeed, to determine whether she was a private merchant ship or perhaps a treasury ship of the Dutch East India Company. Because of weather conditions, it is possible to dive in the Great Basses Reef only in the early spring, so they had to wait at least until the next season before resuming the salvage. By February, 1962, they had bought a sizable boat and a large quantity of salvage equipment. Then, at the end of February, while Clarke was in downtown Colombo buying equipment for the boat, he hit his head a sickening blow against one of the low doorway arches that are common in the Colombo shops. It was not until a few days later, when he was carried, paralyzed and semi-conscious, to a private nursing home, that he learned he had severely injured his spine. At the end of six weeks in the nursing home, Clarke was taken home, where, after some months, he was able to be propped up on the front lawn and begin to write again. During this period, at the rate of a couple of pages a day, he wrote *Dolphin Island,* a novel about the sea. It was also at this time that he received the news that he had been awarded the 1961 Kalinga Prize, given by UNESCO for excellence in science writing. (It had previously been awarded to such men as Bertrand Russell and Julian Huxley.) This was an immense lift to Clarke's morale and, since the prize carried a cash award of £1,000, a lift to his finances as well. In fact, by 1963 he felt both well enough and rich enough to make a full-scale attempt to salvage the treasure ship. His optimism almost cost him his life. He was so weak from his accident that he had to have his aqualung strapped on for him and then had to be pushed into the sea. Underwater, he was caught up by a tidal current that he lacked the strength to swim against, and he was very nearly dead when he made it back to safety. Despite the near loss of its leader, the expedition was successful in salvaging a good deal more silver coin; one of the lumps—a small fraction of the find—was given to the Smithsonian Institution, which has valued it at $2,500. Clarke is sure that a great deal more silver is still buried with the wreck, but he has not taken the trouble to find out. After his 1963 expedition, he was too exhausted emotionally and physically to organize another one, and furthermore, early in 1964, shortly after writing *The Treasure of the Great Reef,* which describes the expedition, he received, out of the blue, a letter from Stanley Kubrick that was the genesis of *2001*.

Clarke is especially fond of Robert Bridges's assertion in *The New Testament of Beauty* that life consists of the "masterful administration of the unforeseen." Thus, when he received Kubrick's letter saying that Kubrick intended to do a film on space and inquiring whether Clarke had any ideas along those lines, he responded immediately and enthusiastically. Clarke had seen Kubrick's *Lolita* and admired it greatly, but had never met him. Kubrick, for his part, having started to soak up space lore, found that Clarke's name was at the end of every avenue of inquiry. As it happened, Clarke at about this time was scheduled to go to New York to work on a Time-Life book called *Man and Space,* so he made arrangements to see Kubrick, who was then living in the city. In New York, Clarke installed himself in the Hotel Chelsea, which is his headquarters whenever he is here. He and Kubrick first met for lunch at Trader Vic's, where they talked for several hours. In the days that followed, Clarke recalls, "I was working at Time-Life during the day and moonlighting with Stanley in the evenings, and as the Time-Life job phased out Stanley phased in. We talked for weeks and weeks—sometimes for ten hours at a time—and we wandered all over New York. We went to the Guggenheim and to Central Park, and even to the World's Fair. We considered and discarded literally hundreds of ideas, which I shall be mining for short stories for years to come. When you read them, you won't have the slightest suspicion that they were connected with *2001.*" In time, the two men settled on a documentary-like depiction of planetary exploration, to which Kubrick gave the half-serious title *How the Solar System Was Won.* Kubrick then suggested that they write a novel on which they could base the eventual screenplay. "That way, Stanley thought, we would generate more ideas, and give the project more body and depth," Clarke explains. Clarke retired to the Hotel Chelsea with an electric typewriter, and for the next several months he and Kubrick produced a couple of thousand words a day. He found this period an extremely happy one, and memorialized it in a little parody of *Kubla Khan:*

For M-G-M did Kubrick, Stan,
A stately astrodome decree,
While Art, the science writer, ran

Through plots incredible to man
In search of solvency.

By 1965, *How the Solar System Was Won* had transformed itself into the enormously ambitious theme of a space odyssey. Almost from the beginning of his study of space literature, Clarke has been struck by the similarities, aesthetic and philosophical, between space voyages and the ancient voyages of the Greeks. He is very much concerned about the fact that there are no geographical frontiers left on the earth to drain off our exploratory energy. He feels that manned space flight, apart from whatever scientific and technological results it will have, provides the necessary new frontier. Five years before Clarke began working on *2001,* he published a collection of essays entitled *The Challenge of the Spaceship,* in which, at the end, he pays a tribute to Homer:

> Across the gulf of centuries the blind smile of Homer is turned upon our age. Along the echoing corridors of time, the roar of the rockets merges now with the creak of wind-taut rigging. For somewhere in the world today, still unconscious of his destiny, walks the boy who will be the first Odysseus of the Age of Space.

While they were working on the film, Clarke once remarked to Kubrick, "If this film can be completely understood, then we will have failed." What he meant was that he and Kubrick were trying to evoke an awareness of an intelligence infinitely old and infinitely different from anything that we can now know or imagine. By definition, such an intelligence could not be understood by us, and its workings, by Clarke's third law, would be to us indistinguishable from magic. Whether or not *2001* has been completely understood, Clarke is delighted with it, and he is fond of quoting a critic who remarked that visually it was so beautiful that "one could take any frame, blow it up, and hang it on the wall."

When in residence in Colombo, Clarke begins his day promptly at seven o'clock with the serving of the morning tea—needless to say, Ceylonese tea. His at-home attire consists of a sarong—the local costume—and sandals. (At the time of my visit, he was favoring a snappy blue number that bore a family resemblance to his living-room curtains.) Thus attired, Clarke transits from his bedroom through Thambi's office and into what he calls his "ego chamber." This is where he works, and it is also a sort of living museum of Clarke memorabilia. It contains the Zeiss microscope; a very elaborate German shortwave radio; two tape recorders, which Clarke uses for dictation; an electric typewriter; a wall of bookshelves filled entirely with Clarke's own works, in all editions (about 300 of them) and all languages. On the other walls are assorted documents and pictures, including the picture of the Stuart Ballantine Medal; a copy of Clarke's 1945 paper on communications satellites, which Mike Wilson had framed and gave to him; several pictures of Clarke with various figures of the space age, including the late Wernher von Braun and the Russian cosmonaut Aleksei Leonov; a picture of Clarke wearing, for some reason, an American Indian war bonnet, and another of him shaking hands with U Thant; a framed letter from Senator Lyndon B. Johnson thanking Clarke for a book, and a similar one from Lady Bird Johnson inviting him to dinner at the White House with the Apollo 8 astronauts. (The astronauts are all Clarke fans, and some of them have told him that their interest in space was originally kindled by his books. Clarke was informed that William Anders, on the Apollo 8 flight, had for a fleeting moment been tempted to report back to earth that he had seen a black rectangular slab on the moon's surface.) Clarke told me that when the ego chamber is fully decorated, the walls will reflect a luminous white ray of pure and undiluted ego—a ray of such intensity that any transient droplet of self-doubt that happens by will be consumed as if by a laser beam.

At seven-twenty, Clarke tunes in to the BBC shortwave reading of editorials from the British dailies, and at seven-thirty, he gets the BBC news. He also keeps in touch with people in New

York and London by telephone, but, having overheard several of these conversations, which are carried out with great shouts of "Yes, I hear you perfectly!" followed by "What did you say?" I should imagine that they leave something to be desired. At about eight o'clock, at least when I was visiting, Carolis would announce breakfast, which was served on a long table in the dining room, and at which Clarke is joined by any of the innumerable houseguests who may be around. Clarke then retires into the ego chamber, where he dictates letters or writes until ten, when the postman arrives on his bicycle, to the accompaniment of a great cacophony of Clarkean and neighborhood dogs. The mail is then sorted out and read, and Clarke dictates any urgent replies into a tape recorder, to be transcribed later. (Clarke is bombarded with fan letters from all over the world, and he has composed a printed form letter for use when he is out of Ceylon and cannot reply personally.) Clarke's afternoons are frequently devoted to reading. In the main, he reads science fiction or books on science. He subscribes to a vast array of popular and technical journals, including the *National Geographic; Scientific American;* the journals of the British Interplanetary Society, the Royal Astronomical Society, the British Astronomical Association, and the American Institute of Aeronautics and Astronautics; *Science,* the journal of the American Association for the Advancement of Science, which he joined; the *New Scientist; Sky & Telescope; Sea Frontiers,* which is an underwater journal; *Time; Newsweek;* and assorted British Sunday papers. The first science fiction that Clarke ever read—in 1929—appeared in an early edition of Hugo Gernsback's *Amazing Stories.* In fact, Gernsback, a Luxembourg-born prophetic genius who spent most of his life in the United States, and died here a few years ago, coined the term "science fiction." The science-fiction writer's equivalent of the Oscar is called the Hugo, and Clarke won it in 1956 for a short story called "The Star." Clarke, who greatly admired Gernsback, dedicated his book *Profiles of the Future* to "Hugo Gernsback, who thought of everything." After *Amazing Stories,* Clarke discovered H. G. Wells, especially *The Time Machine,* which he regards as Wells's masterpiece. Then, at about fourteen, he read Olaf Stapledon's *Last and First Men,* an extraordinary document that covers the next billion years of human evolution, dealing with all the races of man and with all the planets of the solar system. "The book transformed my life," Clarke says.

Science fiction, like science, runs in fads. The first fad was for moon voyages. (*Somnium—The Dream*—by Kepler, the first science-fiction story in the modern sense, was, to be sure, about a moon voyage.) In the early days, these stories dealt largely with elderly professors and their beautiful daughters, who made rockets in their backyards and went off to visit our ancient satellite. "They had plots that we would find laughable today," Clarke has remarked. "Many of the things that excited us then are now reality, and I often wonder if youngsters now discovering science fiction for the first time get the same charge out of it that we did then." I once asked Clarke if modern science had not caught up with, and even surpassed, the imagination of science-fiction writers. He said that this was a common fallacy. "You can't have science fiction until you have some science," he went on. "It wasn't until the astronomical discoveries of the seventeenth century, for example, that one could write at all about travel to the planets. The more the frontier of science expands, the more scope there is for scientific speculation. Certain themes just drop out—like the conquest of the air, which so concerned the science-fiction writers early in this century. We simply move on to more mature and more advanced themes. As long as there is science, there will be science fiction."

Before visiting Ceylon, I did a certain amount of reading about the island—especially in a wonderful, crusty book called *Ceylon: Pearl of the East* by a retired tea planter, Harry Williams. Mr. Williams gives a particularly enticing description of Adam's Peak and the pleasures of climbing it. However, other writers have noted that there is a good season and a bad season for the climb;

during the bad season, apparently, the upper part of the mountain is largely inhabited by leopards and monkeys. As it happened, I was in Ceylon during the good season (the thought kept occurring to me that although I knew that it was the good season, how could I be sure that the leopards knew?), and I asked Clarke if he wanted to climb Adam's Peak with me. He said that he thought it a splendid idea, and that while his old spinal injury might prevent him from getting all the way to the top, he wanted to go as far as he could; indeed, Hector would want to come along, too. We made some inquiries and discovered that the approved method for doing the climb is to leave Colombo at about eight in the evening, drive about four hours to the base of the mountain, and then begin climbing about midnight, when it is cool, arriving at the summit just before sunrise. Clarke made arrangements with a local taxi company for a driver and a car and changed from his sarong into a pair of relatively heavy long pants, a couple of sweaters, and a windbreaker, which made him look vaguely like Robert Morley. The taxi was late, and when it arrived, Hector, knowing the condition of the upcountry roads, insisted on examining its tires, which he pronounced unfit. Clarke then commandeered a second taxi—a shiny new Peugeot with impeccable tires—into which we bundled ourselves, along with an elaborate picnic that had been prepared by Hector's fiancée. The entire Clarkean staff stood by to wish us well, and we drove off into the warm, clear night. We passed through the lowland paddy fields and rubber plantations and moved up into the highland foothills, where the countryside changes from rubber to tea. After about four hours of driving, we caught our first glimpse of the mountains, and while I have seen a good many mountains, the sight of Adam's Peak at night is one of the most extraordinary I have ever encountered. The trail that hundreds of pilgrims take from the tiny town of Maskeliya to the summit is a four-mile illuminated flight of steps reaching 3,000 feet into the air. At night, it gives the impression of some sort of fairyland staircase that has disappeared into the sky, leaving a trail of lights behind.

Our driver stopped in Maskeliya, and we proceeded on foot through the dimly illuminated shrublands. I have a reasonably flammable imagination and began to see leopards emerging from every tree. Clarke said that, considering the hundreds of pilgrims the leopards had to choose from, we would feel pretty foolish at being the meal selected. This point of view carried a certain weight with me until I began to be aware that we were more or less alone on the trail. Later, we learned that most people climb halfway up the mountain on one day, sleep most of the night in one of the rest houses along the way, and then start the final climb just before dawn. After what seemed like a year of climbing stone steps—literally thousands of them—I found that I had pulled ahead of Clarke and Hector. A gong sounded from the Buddhist temple at the summit, and I forged ahead. At the top, I found myself in a large crowd of pilgrims standing in line to see the imprint of the foot of Buddha (or Adam) on the stone in the small shrine at the very summit. It was now about five in the morning, and bitterly cold. I was huddling in a corner of the temple with a large group of freezing pilgrims when Clarke and Hector hove into sight. Clarke had suffered a few bad muscle cramps on the way up, but had made it to the top not too much the worse for wear. Just before sunrise, the Buddhist priests who serve the temple began chanting mantras and playing drums and trumpets, as if to bring the sun into the world by an act of will. It came up over a distant mountain range and illuminated in rosy pastels the whole of the island, which lay spread out beneath us. Sometime soon after sunrise, an almost incredible optical phenomenon took place—one for which the mountain is celebrated. Suddenly, a black triangular shadow appeared in the sky as if by magic—the mountain casts a shadow on the haze, and since the haze is all but transparent, one has the impression that the shadow has emerged from nowhere. At the appearance of the shadow, the priests chanted, the trumpets sounded, and the drums rolled, after which we and the pilgrims started the long trip down. On the way, we met people coming up, and when they saw us they would sing a song, which Hector said was a blessing on all those descending, while the pilgrims with us responded with a song of blessing on those who were climbing to worship at the shrine. Though Ceylon does not have the misery and starvation of its neighbor India, it is a poor country by Western standards. For the pilgrims, climbing Adam's Peak is a religious experience that eases the burden of their daily lives, and a sense of their exhilaration was in their song.

Toward the end of my visit to Ceylon, Clarke suggested that I come with him to inspect what he thought might become the future headquarters of the Far Eastern branch of the Rocket Publishing Company, the Clarke family enterprise that invests Clarke's European royalties. To get to the proposed future branch office required a drive of several hours along the south coast of the island, past the ancient port of Galle, which some scholars think may be the Tarshish of the Bible, and into a tiny seaside fishing village well off the main road. (On the way, Clarke had lapsed into an uncharacteristic total silence, and just before we got to the village he announced that he had worked out the details of a short story involving a man who goes to a fortune teller and is told that he is going to get killed by a prehistoric reptile—a prophecy that comes true a few days later when he visits the American Museum of Natural History and a dinosaur skeleton falls on him.) The future headquarters turned out to be a rather ancient and neglected barn, built by a Dutch settler some two hundred years ago. Clarke felt that he might renovate it and use it as a weekend house or as a base for underwater exploration. At the time, it was largely inhabited by local fauna, including several village cows. Indeed, as Clarke made his way across the future front lawn he flushed a sizable water monitor, which, from stem to stern, was about four feet long. Hitching up his sarong, he lit out after it, and was giving it a reasonably strong chase until he stepped into a nettle patch. Limping slightly, he then led me to the beach, which turned out to be as beautiful a tropical beach as it is possible to imagine. The sand was as fine as sugar, and palm trees fringed the shore. Just beyond a coral reef, several catamarans were lazily sliding out to sea, while a Ceylonese fisherman, perched on a post, was casting his line into the surf. Clarke suggested a walk to the top of a nearby hill, from which we could see the Indian Ocean on almost every side. He said that there was no land between that part of the island and Antarctica, 5,000 miles due south. He then revealed that it was here that he planned to build a house and retire someday. As we were walking back down the hill, Clarke said, "You know, the celestial equator is almost directly overhead. This means that all the communications satellites, in their twenty-four-hour orbits, will be right there, 22,000 miles above my head. I found out quite recently that when a synchronous satellite runs out of its propellant—which keeps it over the right spot on earth—it drifts along this equatorial orbit to two low spots in the earth's gravitational field. One of those low spots is over the Galapagos, and the other one, which is slightly lower, is over Ceylon. So in my old age, if I have a powerful enough telescope, I will be able to look straight up and see some other retired products of the space age—the old and decrepit synchronous satellites sitting almost vertically above me." He added, as we reached the car, "I think there is a kind of piquant symbolism about this. Don't you?"

7

CALCULATORS: SELF-REPLICATIONS

IN the spring of 1955, I completed my Ph.D. thesis for the Harvard Physics Department. I had done a theoretical problem, my principal memory of which is that it required the computation of seventy-five numerical integrals. An integral, for the nonspecialist, is simply the area under a given curve. Each integral took me something like two days to compute; and the seventy-five, nearly six months. Some elementary integrals could be looked up in books,* but mine could not. I had to calculate them myself. The mind boggles now at how I once spent six months of my life—and not unhappily: Sisyphus, as Camus pointed out, was basically a happy man.

Just as I was finishing this task, there appeared on the scene a

*The numerical integrals published then dated back to the Depression, when otherwise unemployed mathematicians were engaged by the WPA in calculating them by hand—that is, with the aid of mechanical calculators. The better old desk calculators could add, subtract, multiply, and divide. Their gears were run by electric motors; each step had to be performed in order, and all intermediate results had to be tabulated. These instruments seem to have taken their place in nostalgia along with the Packard.

whiz kid from Cal Tech. He was about to begin a post-doctoral appointment at MIT, and it turned out, much to my amazement, that he had done essentially the same problem with two notable differences: (1) His basic theory was better, and (2) he had cultivated a group at MIT that had built one of the first electronic computers—a vacuum-tube affair—and, though a dinosaur by present standards, it enabled him to calculate each of his integrals in about two minutes. I decided that in the future I had two choices: I had either to learn to deal with electronic computers, or to avoid numerical integrals. In the past twenty years, I have, by and large, chosen the latter alternative.

I bring all of this up because a number of morals can be drawn about man and the machines he has created. In science, theory and experiment interrelate but, generally speaking, an experimenter begins from some sort of theoretical premise. For example, in high-energy physics, these premises can take such a simple qualitative form as: If the laws of nature were symmetric between particle and antiparticle, then the neutral-pi meson could decay only into an even number of light quanta. Don't worry if that terminology is unfamiliar. The point is that this is a statement that could be—indeed *was*—made without aid of any computer. Even in the case of theoretical statements involving numerical work done by computer, the theoretical structure is always developed by a physicist and not the computer. Computers do not create theories of physics. And no paper in theoretical physics, at least until now, has required the use of a computer to be understood.

In a typical modern high-energy experiment, millions of events occur—photographed tracks in a bubble chamber, or photographed spark discharges, for example. Experimenters make extensive use of computers in recording and analyzing these events. It is not uncommon for an experimenter to prepare a plot on which each point itself represents a million events. Obviously, this kind of data processing must be done by computer. But for the results to be comprehensible, they must be fitted by some sort of curve that arises from a humanly created theory. The chain is human to human, with the machine somewhere in between. It is sometimes said that the computer is to this process as the microscope is to vision. I think this analogy is flawed. Forty-eight people staring at a glass of water with the unaided eye will not see the microbes swimming around in it, while one person with a microscope will see them. The microscope has revealed something that was, in principle, invisible to the human eye. On the other hand, there is nothing the computer does in this chain that could not be done by enough human brain power. Although it took me six months, I did finally do the numerical integrals.

Recently I had the opportunity of discussing this matter with Arthur Clarke, whose fictional computer HAL—invented in collaboration with Stanley Kubrick for *2001*—could do about *anything*. Indeed, HAL could do too *much* of anything. Clarke made the point that electronic computers can now do computation that would require the whole human race working together to accomplish without them (perhaps a more estimable activity than many others the race often engages in). Furthermore, the speed of the individual operations on the computer is something totally beyond human capability; its basic calculations can be completed in the millionths of a second or less. Still, in the man-machine-man process, the machine in the middle can, at least in principle—and, at least, at present—be replaced by humans. Clarke once wrote a story entitled "Into the Comet," in which a spaceship's computer fails and, lacking the proper orbit, the ship heads for disaster. A crewman, of Japanese origin, teaches the entire crew to make abacuses with wires and beads and the ship is saved.

The computer has quantitatively enlarged the sort of calculations and experiments an individual scientist can take on in his lifetime, but, as far as I can tell, it has, on its own, created nothing. In this respect, I am reminded of my one and only encounter with the great Hungarian-born American mathematician John von Neumann. It was von Neumann who developed the theory of stored programming—that is, the capacity of a computer to modify its own instructions as a computation unfolds. He delivered a series of lectures at Harvard while I was an undergraduate there, and I was enormously impressed. After one lecture, I found myself in Harvard Square alongside the great man himself as he hurried by to find the subway. I thought, correctly as it turned out, that this would be the only chance I would ever have to ask him a question. I seized the occasion. "Professor von Neumann," I asked, "will the computer ever replace the human mathematician?" "Sonny, don't worry about it." Literally, that is what he answered.

All this having been said, it must be added that the computer has injected something into modern scientific thinking beyond mere technology. For the first time, I believe, it has presented us with a machine-tooled model—still primitive—of ourselves. I recently read in a history of computers (*The Computer from Pascal to von Neumann,* by Herman Goldstine) that in his early papers on the logical design of computers, von Neumann took his notation from two physiologists, Warren S. McCulloch and Walter Pitts, who were trying to make a mathematical model of the human nervous system. Von Neumann was trying to create what might be described as an electronic nervous system. Now, clearly it would be a fatal mistake to try to construct a model of the nervous system by working from the outside in. That is to say, if one takes as primary data products of the nervous system such as Einstein's theory of relativity, Beethoven's Ninth Symphony, and Van Gogh's *Starry Night* and, from these, attempts to deduce the construction of the apparatus that produced them, one is not likely to get very far. It would be like trying to deduce the structure of the elementary particles of subnuclear physics by contemplating Mount Everest. The idea, rather, is to put together a vast array of very primitive objects and to see what such an array working in concert can produce. The fundamental components of the McCulloch-Pitts model, described in their celebrated paper entitled "A Logical Calculus of the Ideas Immanent in Nervous Activity," were "neurons" connected by wires that could transmit electrical pulses. (In the human brain there are about ten billion neurons—organic molecules about a hundred-thousandth of a centimeter in diameter—wired together by axons, or fibers, that can be several feet long.) For purpose of the analysis, a neuron acts as a relay station for electrical pulses. If such a station receives a sufficiently strong impulse, it will "fire," or emit a pulse. If these neurons are wired together, in units, things can be arranged so that it takes the activation of, for example, a pair of neurons to fire a third one and so on. That event in this so-called logic circuit might be described, in the language of formal logic, as *A plus B implies C.* Now, it has been known since the pioneering work of Bertrand Russell and Alfred North Whitehead that even the most complicated mathematical statements can be broken down to a collection of such primitive logical propositions. Hence, McCulloch and Pitts were emboldened to conclude: "Anything that can be exhaustively and unambiguously described, anything that can be completely and unambiguously put into words is, *ipso facto,* realizable by a suitable finite neural network." In other words, neural networks can carry out the processes of mathematical logic.

Von Neumann was deeply impressed by this analysis, since the computing machines he was designing were essentially neural networks with electronic devices—vacuum tubes in the original, and now archaic, versions of the machines—playing the role of the organic neurons. These machines, therefore, could in principle do anything that a McCulloch-Pitts model could do. Being the kind of genius that he was, von Neumann did not leave the analysis there. He built on the work of a remarkable young British mathematician, Alan Mathison Turing, whose work remains largely unknown, except to specialists. Yet it may turn out that, when a future historian of automation looks back at the really revolutionary implications of the so-called computer revolution, these will have much more to do with the as yet unrealized abstract ideas of Alan Turing, as generalized by von Neumann, than all of the new airline reservation systems, and the like, put together.

Alan Turing, whose life is described in a moving book written by his mother in 1959, died five years earlier at the age of 42, perhaps by suicide. From 1936 through 1938, he studied at Princeton, where his work came to the attention of von Neumann. (Some of the work was done independently by the American logician Emil L. Post, but apparently von Neumann was not aware of it.) Von Neumann offered him a position as his assistant at the Institute for Advanced Study. Turing declined it, preferring to return to King's College, in Cambridge, England, where he was a Fellow. He later worked on the construction of the first British computers.

Turing invented what is now known as the Turing machine, which is actually not a real machine at all but rather, an abstract construct—an idea—for an apparatus that could be instructed to make mathematical calculations. There are various ways of describing the basic idea. But as Mark Kac, a professor of mathematics at Rockefeller University, has put it, the Turing machine has an infinitely long tape divided into identical-sized squares, each one of which either is blank or contains a slash. Over the tape, there is a movable arrow. The machine can be given four basic directions, denoted: L, R, * and /. The L means move the arrow one step to the left; R, one step right; * means erase the slash; / means print slash.

The machine may be programmed to carry out a sequence of operations (see Figure 1). Assume for the sake of the discussion that a given Turing machine had four slashes on an infinitely long tape, one each in Squares 10, 11, 12, and 13. To get it to double those four—in other words, to get it to multiply 4 × 2—one would first move the pointer to Square 9, and then issue a sequence of instructions, bearing in mind that at each step the machine has to be given two alternative courses of action: (1) R (that is, move the pointer one square to the right): if blank repeat

Step 1 (that is, move another square to the right), or if not blank erase / and go to Step 2; (2) L: if blank print / and go to Step 3, or if not blank reprint / and repeat Step 2; (3) L: if blank print / and go to Step 4, or if not blank reprint / and repeat Step 3; (4) R: if blank leave blank and go to Step 1, or if not blank reprint / and repeat Step 4. Now this entire sequence is repeated four times, after which the answer appears as a series of eight slashes, in Squares 5 through 12. (Unfortunately, for the pointer and the machine's imaginary fuel supply, there is no way in this early Turing program to stop the pointer; after completion of the fourth series, it keeps encountering blank squares, leaving them blank and moving one space to the right—on to eternity!)

TURING MACHINE

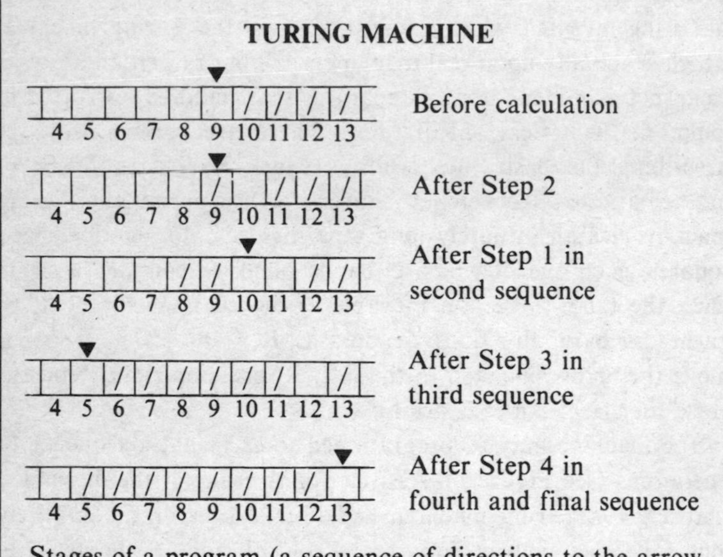

Stages of a program (a sequence of directions to the arrow to move one square right or left and to print a slash or erase one) for multiplying 4 times 2 on a theoretical computer invented in the 1930s by Alan Turing.

Figure 1

All this may seem a bit primitive, but Turing went on to prove a most remarkable theorem: that it is possible to construct a general-purpose machine—he called it a universal machine—on whose tape one can write any number of programs, or codes made of slashes and blanks, and that the universal machine would read these instructions and carry them out. This universal machine can carry out any set of operations that any given Turing machine is able to carry out. It is the abstract embodiment of all Turing machines. (In Turing's time, it was believed that a universal machine would have to be enormous and have to be given millions of instructions, but now the theory has been greatly simplified. Professor Marvin Minsky of MIT holds the record for building the smallest universal machine—it has twenty-eight instructions.) In Professor Kac's words: "The Turing machine owes its fundamental importance to the remarkable theorem that all *concrete* mathematical calculations can be programmed on it. . . . In other words, every concretely stated computational task will be performed by the machine when it is provided with an appropriate, finite set of instructions." (Turing also showed that problems exist for which no program can be devised in principle. These are the computing machine analogues of the undecidable propositions of the mathematics of Gödel. The machine is, in this respect, no better or worse off than the human mathematician.)

Now for the great leap forward. Von Neumann asked himself whether programs could be devised that would instruct the Turing machine to reproduce itself. It had always been supposed that machines were used to produce objects that are less complicated than the machines themselves, that only biological reproduction transmitted total complexity or, indeed, through mutation, increased the complexity. A machine tool, for example, by itself cannot make a machine tool. One must adjoin to it a set of instructions, and these usually take the form of a human operator. Hence, the complete system is the machine tool plus the human operator. Clearly, this system will, under normal circumstances, produce a machine tool minus the human operator—hence vastly less complex.

Von Neumann discussed these matters in the Vanuxem lectures at Princeton in 1953, and since then these talks have acquired an almost legendary character. They were never fully recorded, however, though fragments appeared in a book entitled *The Computer and the Brain,* and they were discussed in 1955 in an article in *Scientific American* by John G. Kemeny, now president of Dartmouth.

When we discuss self-replicating machines we must be clear about the ground rules. In Kemeny's words: "What do we mean by reproduction? If we mean the creation of an object like the original out of nothing, then no machine can reproduce, but neither can a human being. . . . The characteristic feature of the reproduction of life is that the living organism can create a new organism like itself out of inert matter surrounding it.

"If we agree that machines are not alive, and if we insist that the creation of life is an essential feature of reproduction, then we have begged the question. A machine cannot reproduce. So we must reformulate the problem in a way that won't make machine reproduction logically impossible. We must omit the word 'living.' We shall ask that the machine create a new organism like itself out of simple parts contained in the environment."

Von Neumann showed, as early as 1948, that any self-replicating apparatus must necessarily contain the following elements. There must be the raw materials. In his abstract example, these are just squares of paper—"cells"—waiting around to be organized. Then we need the program that supplies instructions. There must be a "factory"—an automaton that follows the instructions and takes the waiting cells and puts them together according to a program. Since we want to end up with a machine that, like the original, contains a blueprint of itself, we must have a duplicator, a sort of Xerox machine that takes any instruction and makes a copy. Finally, we must have a supervisor. Each time the supervisor receives an instruction, it has it copied and then gives it to the factory to be acted on. Hence, once the thing gets going it will duplicate itself, and, indeed, von Neumann produced an abstract model containing some 200,000 cells which theoretically did just this.

Those who have had some education in modern genetic theory may have heard bells going off, or had the sense of *déjà vu,* upon reading the above description. It could apply as well, in abstract outline, to biological reproduction. We are, by now, so used to the idea of computer analogies to biological systems that they may appear obvious. One must keep in mind that they were not obvious at all; indeed, they are only a few decades old. Von Neumann's analysis was five years ahead of the discovery of the double-helix structure of DNA, and preceded by several more years the full unfolding of what is called the "central dogma" of genetic replication. In a Vanuxem lecture in 1970, Freeman Dyson of the Institute for Advanced Study made a sort of glossary translation from von Neumann's machine to its biological counterpart. The "factory" is the ribosomes; the copying machine is the enzymes RNA and DNA polymerase; the supervisor is the repressor and depressor control molecules; and the plan itself is the RNA and DNA. Von Neumann was there first.

His early training in Budapest was as a chemical engineer, and he never lost his feeling for engineering practicalities. He was not content to think purely in the abstract. Hence, he raised the following question: Real automata, including biological ones, are subject to error. There is a risk of failure in each of the basic operations—a wire can come loose. How can one design a system that will be reliable even if the basic operations are not completely reliable? The secret was *redundancy*. Suppose, to take an example from Goldstine, one has three identical machines, each of which makes a long calculation in which each machine makes, on the average, 100 errors. The way to improve reliability is to connect the machines, and require them to agree on one step before they go on to the next. If the system were set up so that once two machines agreed they could set the third at the agreed value and then proceed, then it turns out that the chance of error would be reduced from 1 in 100 to 1 in 33 million! Von Neumann concluded that the central nervous system must be organized redundantly to make it function at a suitable error level. This conclusion also appears to be correct. Von Neumann realized, too, that if the universal Turing machine could be made to reproduce itself, it could evolve. If the program was changed, say, by "mutation," and this change was such that the machine could still reproduce, it would produce an altered offspring.

In Freeman Dyson's words, "Von Neumann believed that the possibility of a universal automaton was ultimately responsible for the possibility of indefinitely continued biological evolution. In evolving from simpler to more complex organisms, you do not have to redesign the basic biochemical machinery as you go along. You have only to modify and extend the genetic instructions. . . . Everything we have learned about evolution since 1948 tends to confirm that von Neumann was right."

Where does all of this leave us? The point I have been trying to make is that, as far as I can see, the most profound impact of the computer on society may not be as much in what it can do in practice, impressive though this is, as in what the machine *is* in theory—and has less to do with its capacity as calculator than with its capacity for self-replication.

Recently, I discussed these matters with Professor Minsky, who is one of the foremost authorities on machine intelligence. He told me that von Neumann's complex arguments have now been greatly simplified by his successors. Abstract models of self-reproducing machines have been devised that are extremely simple. Moreover, real computing machines have elements in their design that are beginning to resemble self-replication. One uses a computer to program the design of a computer, and this design is given to a computer that supervises the actual physical construction of the new computer. One must supply from the outside the actual silicon chips on which the circuitry is printed, so, in this sense, the process is not really self-contained. Minsky says most people now agree that a truly self-replicating automaton would have to be the size of a factory (one of whose functions would be the manufacture of silicon chips). It now seems conceivable—in principle, at least, the logic is there—that perhaps with the addition of some primitive biological components (who knows what!), the process can be further developed to the stage where self-producing automatons can be made that are compact and, acting in concert, can do just about anything. In his Vanuxem lecture, Dyson gives several examples of what colonies of these machines might accomplish, for good or evil, if let loose on earth or in outer space—such as bringing vegetation, light, and heat to Mars. With a little thought the reader can supply his own examples. For some reason, as admiring as I am of the logic of automatons, I find the prospect chilling.

I suspect—and this is also emphasized in Dyson's lecture—that for self-reproducing machines to do anything interesting they must have a very high level of interorganization. As Dyson put it: "The fully developed colony must be as well-coordinated as the cells of a bird. There must be automata with specialized functions corresponding to muscle, liver, and nerve cell. There must be high-quality sense organs and a central battery of computers performing the functions of a brain," which may mutate and proliferate. In time, we may no longer recognize them. In this respect, Sara Turing quotes a letter, in her book about her son, that she received from the wife of one of Turing's closest colleagues, M. H. A. Newman. Mrs. Newman wrote: "I remember sitting in our garden at Bowdon about 1949 while Alan and my husband discussed the machine and its future activities. I couldn't take part in the discussion and it was one of many that had passed over my head, but suddenly my ear picked up a remark which sent a shiver down my back. Alan said reflectively, 'I suppose, when it gets to that stage, we shan't know how it does it.' "

8

INNOVATORS: GÖDEL'S THEOREM

For many the abstract is a source of weariness—for me on good days an intoxication and a feast . . . .
Nietzsche

APRIL 28, 1906 *was* a good day. Soft spring breezes wafted over the River Cam. Stem-bent daffodils assisted the chorus of earliest birds. Bertrand Arthur William Russell, the third earl, had just descended from his railway carriage in the Cambridge Station. His step was brisk as he walked along St. Andrews Street. The Great Gate of Trinity College was not far away. He was looking forward to a discussion of ethics with his colleague G. E. Moore. His work with Whitehead had been going well—they were writing the *Principia Mathematica.* The title stirs the blood. Newton's *Philosophiae Naturalis Principia Mathematica* created theoretical physics. Their book, so they thought, would create the foundations of mathematics. All of mathematics, they felt, would be shown to be derivable from a few self-consistent axioms, a few symbols, and a few rules of inference. It would be a logician's dream come true. Russell's sense of well-being was nearly total. Almost accidentally, his left hand swept over his chin. He was temporarily disconcerted by a disagreeable sensation of stubble. In his eagerness to come up from London, he had forgotten to shave.

"No matter," he thought, "I shall visit my barber—Barrett," whose shop was located on the corner of St. Andrews and Market streets. " 'Barrett the barber' belongs to the class of all men whose professions have the same first initial as their last names," thought Russell. " 'Locke the logician,' 'Plato the philosopher,' 'Maxwell the mathematical physicist'—they're all members of the class," Russell noted. He stopped in front of Barrett's shop. Before entering it he looked in the corner of the shop window for the familiar hand-lettered sign. It was there—faded but still legible. It read: "Barrett is willing to shave all, and only, men unwilling to shave themselves."

Russell chuckled softly to himself. The sign had first appeared in the summer of 1902. That June, Russell had written a note to his German colleague, the logician Gotlob Frege. After reading Frege's *Grundgesetze der Arithmetik* (*The Foundations of Arithmetic*) Russell had found an absolutely fatal flaw in it. Frege's definition of "class" was faulty. The class of all classes that are not members of themselves is *not* defined.

Think about it. Turn it over in your mind. If this class *is* a member of itself then, by definition, it is not a member of itself, and if it *is not* a member of itself then it is a member of itself. One is awash in a paradox. Frege understood this almost at once. He wrote to Russell, "Your discovery of the contradiction caused me the greatest surprise, and I would almost say consternation, since it has shaken the basis on which I intended to build arithmetic. . . ." Russell and Whitehead had worked for years afterward to eradicate the contradiction. A "theory of types" had been formulated and a notion of "meta-mathematics" had been introduced. All was now secure—or so they thought.

As Russell entered Barrett's shop, his nostrils were assailed by the smell of perfumed shaving tonics and bubbling soaps. Barrett greeted him with pleasure.

"Ah, Your Grace," he said. "I see that you have come up from London."

"Yes, Barrett," Russell replied. "And *I* see that the old sign is still there."

Barrett braced himself for the inevitable question. "Tell me, Barrett," Russell went on, "in view of your sign, are you willing to shave yourself?"

Barrett flinched.

On this same day, a thousand miles away in the town of Brünn, in the Austro-Hungarian Empire—the same town, by the way, in which Gregor Mendel was born—Kurt Gödel was born.

This was to be the start of something big.

• • •

The phone is ringing in my living room. It is W. W is an idea whose time may be passing—or maybe not. It depends on how you feel about computing machines. W is a human computing prodigal. If you ask him, say, for the fifteenth root of 5,946,-751,320, he will give you the answer—4.483515256—in less than a minute, all done in his head. However, my pocket calculator, which cost me seventy dollars, does the same thing in less than two seconds. I know, because I have just watched it do it. Knowing W, this makes me sort of sad. W has a "trick" for doing roots. God knows what my calculator does. It is hard to work up much interest. W's trick is the following. He has memorized the entire table of natural logarithms of fourteen decimal places—all the numbers from 1 to 100 in steps of a tenth; a thousand, fourteen-place numbers. Some trick. When he was a child, a well-meaning teacher showed him a table of logarithms. W felt that this meant he should memorize it, which he did in about a half hour. He was then ready for the next assignment. When W's teacher found out about this he tried—for W's own good—to interest him in something else, anything else. W couldn't understand what was bothering the teacher. He kept memorizing anything with numbers—the telephone directory, for example. He used to earn his living by giving exhibitions. Now he is retired, but he amuses himself, and his friends, by finding the prime factors in large numbers. For example,

$$9{,}827{,}489 = 7^2 \times 131 \times 1531$$

The numbers 7, 131, and 1531 are all prime numbers. The only thing that divides evenly into them are the numbers themselves and 1, which divides into anything. I used to have to multiply out these prime factors by hand to check W. Now, I can do it on my machine. Arithmetic isn't what it used to be. The time may be approaching when only a rugged individualist will know the ordinary multiplication tables by heart; this is a little like making your own clothes. There is no telling what will happen these days.

"Did you know," W says over the phone, "that your telephone number is a prime?"

"That's very interesting," I answer. "Does that include the area code?"

There is a pause.

"I did not take the area code into account," W notes. "I will have to call you back."

In the first quarter of the present century a notable effort was made to put the foundations of mathematics on secure logical foundations, and although few persons were able to follow the recondite processes of reasoning required it was generally accepted in philosophical circles that the theorems of mathematics could all be deduced from a set of axioms with the sole help of principles of logic. In 1931... Herr Kurt Gödel, then only twenty-five ... challenged this belief. Though Gödel's Proof is even more abstruse than the beliefs it calls in question it has convinced those who are able to follow it.... [Times (London) Literary Supplement]

I like that "Herr Kurt Gödel...." It has a nice democratic sonority about it. We also have, now that we are on the subject,

Herr Albert Einstein, Herr David Hilbert, and Herr Immanuel Kant. They will also play a role. But for the time being, we are all Herrs together. But to get back to our story.

In the 1930s, Gödel and Einstein both became associated with the Institute for Advanced Study in Princeton (thanks, of course, to that master logician Herr Adolf Hitler); it was only the international reputation of the two men—and the immigration policies of the United States—that saved them from becoming lamp shades or bars of soap. In due course, I have been told, the two men became good friends. Gödel, it appears, used to accompany Einstein on Einstein's daily walk home from the Institute to his house at 112 Mercer Street. What I would give to have been with them. There would have been a language problem, since they liked to speak German to each other, and I don't know that much German, but perhaps something could have been worked out.

I can visualize, I think, how those walks must have gone. Sometime after tea—the afternoon tea at the Institute is served at about four—Gödel might have knocked at the door of Einstein's office. Greetings might then have been exchanged, and figure-laden notebook papers collected and put into briefcases. The two of them would have made a contrasting pair as they walked down the front steps of the Georgian office building onto the drive—Einstein a large massive man and Gödel spare, slight, and intense. They would walk down Olden Lane, past the tree-shaded houses of the full professors of philosophy, and onto Mercer Street. Here they would have turned right. Some minutes later they would have arrived at Einstein's house, still talking, perhaps, about the foundations of quantum mechanics or of space and time. Gödel, one gathers from his writing, had come to believe, like Kant and Parmenides, that "change" was an illusion or an appearance "due to our special mode of perception"—an arresting idea—and that, moreover, Einstein's theory of relativity provided "unequivocal proof" that this was so. These are deep waters. What I have always marveled at is, that out of these purely philosophical ideas, Gödel produced a new set of solutions to the equations of Einstein's theory of gravitation. Einstein commented, "Such cosmological solutions of the gravitation equations have been found by Mr. Gödel. It will be interesting to weigh whether these are not to be excluded on physical grounds."

I am interrupted again by the phone, just as I thought I might be getting somewhere. Perhaps it is W calling back with the prime factors of my entire telephone number, including the area code. It turns out, however, to be K. I am very fond of K. I may even be in love with her. But at the moment I am obsessed with Gödel's theorem. K is worried about me.

"I am worried about you," she says. "I haven't heard from you for days. Where have you been? Is anything wrong?"

"Nothing is wrong, at least in the usual sense in which the word 'wrong' is used," I try to explain. "I have been working night and day trying to understand the proof of Gödel's theorem. I can't help it, but that is all I seem to be interested in right now."

"But I am interested in *you,*" K says.

This raises a point in elementary logic—what the logic textbooks call "transitivity." If I am interested in the proof of Gödel's theorem and K is interested in me, then why is K not interested in the proof of Gödel's theorem? If A implies B and B implies C, then, one would have thought, A implies C. In this case, transitivity seems to have temporarily broken down.

"How can I be interested in the proof of Gödel's theorem if I don't understand it?" K says, very reasonably. The quandary that I find myself in is that I have become obsessed by the proof of Gödel's theorem precisely because I *don't* understand it, and, furthermore, I don't understand *why* I don't understand it. Does that make sense?

Gödel has certainly tried to be helpful. Take the matter of translation. The great papers of 1931 are in German. But Gödel read and approved Jean van Heijenoort's English translation. "Professor Gödel," van Heijenoort writes, "approved the translation, which in many places was accommodated to his wishes. He suggested, in particular, the various phrases used to render the word *'inhaltlich.'* He also proposed a number of short interpolations to help the reader and these have been introduced in the text below between square brackets."

In the translation of Theorem IV of the *"Einige metamathematische Resultate uber Entscheidungsdefinitheit und Widerspruchsfreiheit,"* to take an example, one finds:

> Theorem I still holds for all—consistent extensions of the system S that are obtained by the addition of *infinitely many* axioms, provided the added class of axioms is decidable [[*entscheidungsdefinit*]], that is provided it is metamathematically decidable [[*entscheidbar*]] for every formula whether it is an axiom or not (here again we suppose that the logic used in metamathematics is that of *Principia Mathematica*)."

The square brackets [[*entscheidungsdefinit*]] and [[*entscheidbar*]] are, one gathers, due to Gödel.

Take the phrase *"Entscheidungsdefinitheit und Widerspruchsfreiheit."* It rattles around in one's mind like a chorus of French horns. One can look up each word in the dictionary. One can look up *all* the words in the paper in the dictionary, for that matter. *"Entscheidungsdefinitheit und Widerspruchsfreiheit"*—"Completeness and Consistency."

The meaning of each word in the paper added together is, alas, not the meaning of the paper, any more than the meaning of each word in a poem adds up to the poem taken as a whole, or any more than each separate note of a symphony is the symphony. Imagine reciting one of Shakespeare's sonnets by reading one word a week. After one had done this, would one have recited the sonnet? And if not, why not? What has been left out? But I am getting off the subject again.

The phone is ringing. This time it *is* W. He has found the prime factorization of my entire phone number, including the area code. I note it down on a slip of paper. I want to verify it later on my pocket calculator. There is a long pause in the conversation. Perhaps W would like another number to factorize. I give him my Social Security number, which has nine digits. A nine-digit number is child's play to W, once he puts his mind to it.

"I will call you back tomorrow," he says. "Tonight I am going to the movies."

• • •

In Nepal, I once listened to a group of Tibetan monks chanting. They had trained themselves—somehow—so that each of them could chant more than one note at a time, like a violin. A single monk sounded like a small choir. He was able to vocalize a single note and its harmonics all at once. The effect was like nothing else I have heard on this earth.

Perhaps to understand the proof of Gödel's theorem one needs a mind trained so that it can hold all of the strains of the argument together in a single idea. The idea and all its harmonics, all its resonances, would appear in one's mind simultaneously. Great mathematicians, it appears, "see" the entire theorem *before* they construct the formal argument. The argument proceeds in steps, but, in the original, illumination occurs in a single step.

Take Gödel's paper of 1931—*"Über formal unentscheidbare Sätze der Principia mathematica und verwandte Systeme"* ("On formally undecidable propositions of *Principia Mathematica* and related systems"). It begins with forty-six preliminary definitions. Definition 1, for example, is the definition in the language of the *Principia Mathematica* for the process of the division of one whole number by another. It reads:

"$x/y = (Ez)\ z \leq x\ \&\ x = y.z$, x is divisible by y"

To say it in words:

"If x is divisible by y that means that there exists a number z less than x such that x is equal to y multiplied by z."

There are forty-five more definitions like this—and of increasing complexity. Now, and this is the question; Did Gödel know from the start that he would require forty-six preliminary definitions in order to prove his theorem and not forty-seven or forty-three? Did Beethoven know from the start that it would take him "n" notes—however many there are—to write the Ninth Symphony? Did Picasso know from the start that he would need "m" brush strokes to paint *Guernica;* and, now that we're on the subject, did Shakespeare know in advance he would need "l" words to write *Hamlet?*

In the beginning there was the idea—the whole thing, the indivisible concept. The notes, the strokes, the words, the forty-six preliminary definitions came later. Therefore, in a certain sense, I can never understand Gödel's theorem. To understand it—to really possess it in my mind—I would have to have *invented* it. Perhaps I can *learn* Gödel's theorem.

"Let us begin the hour by considering the notion of 'paradox'!" I sometimes lecture myself like this when I am trying to understand something. If K or W were here I might try to lecture them. Or, sometimes, I make up allegorical stories and tell them to myself—things like Barrett the barber. Now *there* is a paradox.

"Barrett is willing to shave all, and only, men unwilling to shave themselves."

I can visualize gown-clad Cambridge undergraduates and dons walking down Market Street. Some will glance into the window of Barrett's shop. A few will see the sign and some may even bother to read it. Most—nearly all—of those who do will think to themselves, "That certainly sounds reasonable," and walk on. Now, in all that human flux, there will be someone with a slight irregularity in the neuronal formation, an extra connection, or maybe, for all I know, a million extra connections—W must have at least a million extra connections—in the neuronal switching centers; and that person will think to say, "But what about Barrett? Who then can shave poor old Barrett?" If Barrett is willing to shave himself, and if his sign is true, then he is not willing to shave himself. And, furthermore, if Barrett is not willing to shave himself, then, according to his sign, he is willing to shave himself. This sounds like the sort of thing that Russell was writing to Frege about. Well, *that* at least we *can* account for, since the man who wrote to Frege—Russell—also invented the "barber paradox." Since I am making this story up, I can also imagine Russell saying to Barrett, "Look, Barrett, there *is* something wrong with your sign. What you need is another sign to go along with it—a 'meta-sign'—which would address itself to the first sign. It might read, 'Barrett should be excluded from the class of all men to whom the first sign refers.'" Barrett must be allowed to shave himself, or, at least, to have someone else shave him. "For God's sake, Your Grace," Barrett might say, "hold your tongue and let me shave!"

In really difficult cases, we might have to provide and entire hierarchy of signs—dozens or hundreds of them—each providing warnings and clues as to how to read the others—signs, meta-signs, meta-meta-signs, and so on. Hopefully, this process would converge and the entire hierarchy of signs would be, taken together, a consistent, paradox-free entity. This, one gathers, is what Russell was about when he created his "theory of types." That way Frege's proof of the consistency of arithmetic could be saved. Well, don't be too sure. There is Gödel, just born in Brünn and still to be heard from.

The phone again—it is K.

"When was the last time you had something to eat?" she asks.

"I am not sure," I answer truthfully.

"What was it?" she asks, with a note of concern in her voice.

"I had a can of tuna fish. It must have been a few hours ago." The sun is setting.

"I am coming right over there," K says, "with something that I am going to cook for you."

"All right." I say. "I would appreciate that."

K has arrived. Her face is bright red from the cold. She is wearing a ski cap that fits over her long blonde hair. She is carrying a shopping bag full of groceries. A stalk of something that looks like it might be celery is visible. She looks wonderful. She looks around my living room-study.

"This place is a god-awful mess," she says.

She is right. I have strewn, over chairs and tables, books and notebooks, which I have been using in order to digest the contents of the books. I have forgotten to turn off my pocket calculator and it is glowing an eerie red. For some reason that I can no longer remember, there is a table of logarithms next to the telephone. There are also scraps of paper with W's prime factorizations written on them. For example, $29{,}645{,}000 = 2^3 \times 5^4 \times 7^2 \times 11^2 = 8 \times 625 \times 49 \times 121$. I no longer have the remotest idea to what the number 29,645,000 refers. Perhaps it is the population of some European country.

K has located the empty tuna fish can. It was under the telephone table. I must have been eating the tuna fish while I was talking on the phone.

These matters are entirely appropriate to put on the agenda. Indeed, they form the basis for discussion, and, in fact, a settlement has been reached. It consists of the following elements:

1. All my books and papers are to be collected and arranged somewhere in the vicinity of my desk.

2. Whatever I do in this area is to be regarded as *"my* problem."

3. K is free to do with the rest of the apartment as she sees fit. Heavy lifting, if any, is to be shared.

The first thing to go is the empty tuna fish can. K has also agreed to listen to a lecture, after dinner, on the history of the foundations of mathematics and the role of Gödel's theorem. This lecture, which should be in non-technical language and as free of allegory as decency allows, is to last no longer than a half hour. I am to stay out of her way while she is preparing dinner. I think that I will use this time to prepare the lecture.

K's lecture:

"Mathematics began as an experimental science. Before there was anything that we would think of as abstract mathematics, people were measuring things. 'Geometry,' for example, means 'earth measure.' It must have seemed to them that it was a purely empirical proposition that when one draws a triangle in the sand and adds up the sum of the angles it comes out to be exactly 180 degrees every time; well, not quite exactly, since the surface of the earth is a sphere. But it is such a big sphere that this hardly matters when one is considering a small triangle. Around the fifth century B.C. what we recognize as modern mathematics—Western mathematics—began in Greece. The exact sequence of events that brought this about is not entirely clear, at least to me. But it happened, and the distinction between an experimental science like physics and mathematics began to emerge. From this point of view, the fact that a three-sided, closed figure made up of straight lines—a triangle—has an angle sum of 180 degrees is no longer to be regarded as something to be demonstrated experimentally. It is contained in the definition of what is meant by a straight line. The logic of all of this was first codified by Euclid late in the fourth century B.C. It was an enormous achievement. He created the notion of mathematics as an organized discipline in the same way that Newton created the idea of theoretical physics.

"In his *Elements,* Euclid separated geometry into its component parts. First there were the 'definitions'—'A line is a length without breadth,' for example. Then came the 'axioms,' propositions about the definitions that werc supposed to be accepted without proof—for example, 'Through a point external to a given line one and only one line can be drawn that is parallel to the given line.' Finally, there came the 'theorems,' propositions—like the fact that a triangle has an angle sum of 180 degrees—that were to be proved from the axioms. The laws of logic to be used in constructing these proofs were not very clearly specified, but, such as they were, they had been codified by philosophers like Aristotle and others. Despite the imperfections and lapses of clarity, Euclid created the model of what a deductive mathematical system is, and we have been using this model ever since."

This much I think K will have no difficulty with. She enjoyed geometry in school. That is one of the troubles—"geometry in school" is a "subject." It hardly occurs to anyone to ask whether Euclid's geometry makes sense. How do we know that the whole system makes sense? Perhaps the axioms are not consistent. If we work at it long enough, maybe we can derive the fact that the angle sum of a triangle both is 180 degrees and is not 180 degrees—a flat-out inconsistency. If the axioms are inconsistent, then anything—any goddamn thing—can be derived from them. That is something that they don't teach you in school, or at least not in my school. Anyway, back to K's lecture.

"Here," I will say to K, "are, as far as I can see, the key elements that finally lead to Gödel. First there came what may be called the 'arithmetization' of all of mathematics. This is a long story, but it goes something like this. When you first study it, geometry does not seem to have any direct connection to arithmetic or algebra. In school, these are taught as different 'subjects.' But early in the seventeenth century, Descartes discovered what is called 'analytic geometry.' Euclid's lines and curves are, he noted, the solutions of algebraic equations, and, in fact, all of geometry can be formulated in terms of algebra. But algebraic equations are relations among numbers so that, putting it baldly, algebra and geometry can be reduced to arithmetic. But what is arithmetic?

"Arithmetic begins with the integers, the whole numbers—1, 2, 3, and so on. This is all there is so long as you stick to multiplication, addition, and subtraction. If you do any of these things to two whole numbers you get a third whole number. Fine.

But what about division? By definition, if you like, 1 divided by 3 is not a whole number. It is a new thing, which we call a 'fraction.' Very good. But are there any kinds of numbers that are not simply the quotients of whole numbers? Yes. In fact, the ancient Greeks realized that the square root of 2, to take an example, cannot be expressed as one integer divided by another. It is a new kind of thing, which is called an 'irrational number.' Most numbers, it turns out, are irrational. But we can always find fractions that are as close to any of these irrational numbers as we want. This is getting a little sticky. But imagine that the irrational number is represented as a point on a line. Then I can find points representing fractions that are as close to this point as I like. In fact, I can define the irrational number by choosing fractions approaching closer and closer and letting the limit *be* the irrational number.

"I go into this because I want to impress on you how much of the logical weight of the whole edifice is bearing down on arithmetic—the whole numbers, then the fractions, then the irrationals, then algebra, then geometry, then everything. Arithmetic seems so simple because we learn it even *before* we go to school. But does *it* make sense? Is it a consistent system?

"This question began to be addressed in earnest in the nineteenth century by mathematicians like Peano and Frege, who wrote down what they believed were the complete and consistent axioms for arithmetic. The spirit was the same as Euclid's except that, by then, a good deal of progress had been made in formulating and sharpening the rules of logical inference. What is known as 'symbolic logic' was being created. This is another complicated story, but it reached a sort of climax when Russell and Whitehead produced the *Principia* during the years 1910 to 1913. They showed explicitly how, starting from the axioms, definitions, rules of inference, and so on, one could build up all of mathematics from arithmetic.

"But their work left the big question unanswered. Are the axioms of arithmetic consistent and complete? At about this time, the great German mathematician David Hilbert, and his collaborators, launched an all-out attack on proving the 'absolute' consistency of mathematics and, in particular, of arithmetic. Up to this time mathematicians had only attempted to prove the 'relative' consistency of one system with respect to another. In other words, if one wanted to prove that geometry was consistent, one would find a translation of geometry into arithmetic—an arithmetical model of geometry—so that one could conclude that geometry was as consistent as arithmetic, no more and no less. The best that Hilbert and his school were able to do was to show that a truncated form of arithmetic which had addition but not multiplication was absolutely consistent. Once multiplication was added the whole question of the absolute consistency was up in the air again."

K is stirring something. She is bent over my small stove, her body curved into a graceful arch. Passion—*that's* something to think about. We have no trouble imagining two men killing each other over the love of a woman, or people killing each other over a religious or political ideal. All of this seems quite reasonable—books have been written about it, to say nothing of songs and poems. I have never read a poem about two men killing each other over a mathematical theorem, but it could happen. The passion aroused in disputes over abstractions can be as violent as anything you'd ever want to see. Take the dispute between Hilbert and the Dutch mathematician L. E. J. Brouwer. Hilbert and his school were known as the "formalists," since they insisted on using the formal rules of symbolic logic throughout, while Brouwer and his people became known as "intuitionists." (In an opera, the "formalists" would be all trumpets while the "intuitionists" might be flutes and harps.) The intuitionists refused to accept the laws of symbolic logic in situations where, in a manner of speaking, they could not count up the cases one by one. They liked to get ahold of things with their ten fingers and ten toes. Take the law of the excluded middle, for example. It states that a well-formulated proposition is either true or it isn't. There is nothing in the middle. If you are a formalist, you simply accept this without a struggle. But if you are an intuitionist, you want to be shown. If the proposition refers to an *infinite* number of cases, how are you going to show an intuitionist that the law of the excluded middle applies to all the cases? Well, you aren't. It has to be taken as one of the formal rules of the game. Once one refuses to accept this, school is out and almost anything is possible. In due course, Brouwer and his people, and Hilbert and his people, were hardly speaking to each other. In another age, they might have fought duels. Hilbert's people kept plugging away with their formal methods, trying to prove the absolute consistency of mathematics; while Brouwer and his people kept insisting that the whole program was nonsense, since the usual laws of logic were not applicable to sets with an infinite number of objects. So long as Hilbert's people could not come up with a proof, there was no way of refuting Brouwer, and, even if they had come up with a proof, Brouwer probably would have refused to accept it. Meanwhile, millions of schoolchildren were learning arithmetic with no idea of how interesting and difficult a "subject" it really was, and is.

Back to K's lecture.

"In the late 1920s, when Gödel was a student in Vienna, all of this controversy in the foundations of mathematics was swirling around. I would imagine that a student of mathematics at that time would have found it irresistibly fascinating, just like the then students of physics found themselves caught up in the new quantum theory. It is clear from the references in Gödel's paper that he had immersed himself in the literature on the subject. He knew all about what Hilbert and his people were trying to do. But something must have triggered a doubt in his mind. Perhaps he had tried to prove absolute consistency and had not been able to. He certainly accepted the formalist point of view, in the sense that in all his work he operates within the rules of the *Principia*. These are the rules and he did not try to change them. But using the rules he was able to shift the discourse completely. To put the matter starkly—and I will come back to fill it out—he showed that Hilbert's question was *unanswerable! There is no proof of the absolute consistency of mathematics and no such proof can ever be given.* The question is *undecidable.* Think what this means. At the very center of the most pristinely logical of all the creations of the human mind, there is a question that can never be answered—namely, the consistency of the enterprise itself. I find this to be one of the most ironic dilemmas the human mind has managed to create for itself.

"What is known loosely as 'Gödel's theorem'—in his paper there are several theorems—is the precise demonstration of this fact. What he did was to exhibit a statement in the language of arithmetic—it is a very complex statement—which has the following properties:

"1. The statement is *true.* By *true* he meant that if this statement were tried out on any particular example, one would get a true statement in arithmetic. That he could guarantee.

But

2. The truth of this statement was *unprovable.* No possible proof of this statement could be given in the language of arithmetic itself. Gödel called statements like this 'formally undecidable,' since no method of formal proof in the language of arithmetic could be found, in principle, to prove the statement. One might add this statement to the system as a new axiom, but then Gödel could show that the enlarged system would also have 'formally undecidable' statements, and so one is trapped in an infinite regression.

3. The statement that the system was consistent was itself such an undecidable proposition. That means that Hilbert's question was unanswerable.

"One can imagine the effect that this had on Gödel's contemporaries. Gödel's proof was studied, and no flaw was found. In fact, Gödel had planned to write a second paper to enlarge on the first one, which he had labeled 'I.' Later he wrote, 'The author's intention was to publish this sequel in the next volume of the

Monatshefte. The prompt acceptance of his results was one of the reasons that made him change his plan.' An intuitionist could still deny the validity of the theorem. But then, where is one? What are the rules of the game? Mathematics becomes a form of blank verse.

"About the same time, Heisenberg discovered the Uncertainty Principle in quantum physics, which defines the limitations of physical measurements—it limits the possible questions one can ask about atomic systems. The two ideas do not really have any logical connection, but the spirit is similar. Some questions which language allows one to ask turn out to be meaningless. I often wonder if some of the questions that we are now asking really have any meaning. We are now trying to understand the human mind. We want to make a model of memory, of reason, and all the rest. We are trying to examine ourselves the way a technician examines a computer. Is this really possible? Perhaps it is—but, perhaps it is not.

"Once, at the very end of his life, I went to visit Schrödinger in Vienna. Schrödinger was one of the founders of the quantum theory and one of the deepest scientific and philosophical thinkers of the twentieth century. When I saw him, he was a charming, frail old man. He was very farsighted, and when he looked at you through his glasses they magnified his eyes so that they looked enormous. I went to see him with another young scientist. Schrödinger studied us from above a book-strewn desk, and then he said, 'There is something that the ancient Greek scientists knew that we seem to have forgotten.' There was a pause—a long pause—and then he said, 'and that is modesty.' "

This is about as far as I can take K on this trip. The rest of the voyage I will have to make alone. It was good to tell her about it. She has now gone to sleep in the next room. I am too involved in my own thoughts to sleep. I am at my desk. The scrap of paper with W's prime factorization

$$29{,}645{,}000 = 2^3 \times 5^4 \times 7^2 \times 11^2$$

has just fluttered to the floor. I pick it up and study it. There is one significant thing to say about it. Once W has located the prime factors, they are unique. There is no other way of writing the same number in terms of its prime factors. This is one of the classic results of number theory, and goes under the name of the "fundamental theorem of arithmetic." What I have now come to understand is that these unique prime factorizations are at the heart of the proof of Gödel's theorem. This should please W when I have a chance to tell him.

The problem that confronted Gödel is already present in Russell's barber paradox—Barrett again. In this paradox, the question "Does Barrett shave himself?" is undecidable within the limited language he has chosen to use. We need a new sign to tell us how to read the original one, and this sign takes us out of the original domain of discourse. One might imagine that the same sort of thing would happen if one deals with the provability of a mathematical theorem. This looks, at first sight, as if it is not a mathematical statement at all, but is rather a statement *about* a mathematical statement—its provability. Hence, how is such a statement to be included in the same domain of discourse as the mathematical theorems themselves? Here, Gödel had an extremely beautiful idea, and here is where the prime factorizations come in.

Gödel was able to attach to each statement in mathematics, to each sign, to each symbol, to each proof, a unique number. This is built up step by step. He begins with the fundamental symbols which in the *Principia* are things like

"$\widetilde{n}$" meaning "not"

It turns out that there are seven of them, so each gets a number: 1 to 7. Next there are the variables. If a variable stands for some possible number, Gödel assigned it one of the prime numbers; if it stood for a whole sentence, he assigned it the square of a prime number; and so on. When one puts these sentences together one gets a string of prime numbers and powers of prime numbers which correspond to one single unique number—just like W's prime factorization. Hence, every statement in the system corresponds to some number and all proofs, and all statements about proofs become relations among these numbers. The arithmetic system has been trapped into describing itself in its own language. It is locked into its own domain of discourse, and, as Gödel showed, this leads to the existence of the undecidable statements. We cannot provide additional signs, as we did for Barrett, because these additional signs get caught up in the arithmetic discourse. There is no place to go.

This is the general outline. But the details, the details! The whole thing is becoming clearer to me, but God am I tired! By now the sky is beginning to brighten and the night people have faded away on the street below. I feel serene—almost indescribably happy. It has been a good day.

SELECTED BIBLIOGRAPHY

Armitage, Angus. *John Kepler*. New York: Roy Publishers, Inc., 1967.

Clarke, Arthur C. *Childhood's End*. New York: Harcourt, Brace and World, 1963.

———. *The City and the Stars*. New York: Harcourt, Brace and World, 1956.

———. *A Fall of Moondust*. New York: Harcourt, Brace and World, 1961.

———. *Profiles of the Future*. New York: Harper & Row, 1962.

———. *The Promise of Space*. New York: Harper & Row, 1968.

———. *The Sands of Mars*. New York: Gnome Press, 1951.

———. *The Treasure of the Great Reef*. New York: Harper & Row, 1964.

Koestler, Arthur. *The Sleepwalkers*. New York: The Universal Library, Grosset & Dunlap, 1963.

Lear, John. *Kepler's Dream: With the Full Text and Notes of Somnium Sive Astronomia Lunaris, Joannis Kepleri*, trans. Patricia F. Kirkwood. Berkeley, Calif.: University of California Press, 1965.

Medvedev, Zhores A. *The Rise and Fall of T. D. Lysenko*, trans. I. Michael Lerner. New York: Columbia University Press, 1969.

Olby, Robert. *The Path to the Double Helix*. Seattle, Wash.: University of Washington Press, 1975.

Rabi, I. I. *Science: The Center of Culture*. New York: The New American Library, 1970.

Sayre, Anne. *Rosalind Franklin & DNA*. New York: W. W. Norton, 1975.

Thomas, Lewis. *Lives of a Cell: Notes of a Biology Watcher*. Viking Press, 1974.

Watson, James D. *The Double Helix: Being a Personal Account of the Discovery of the Structure of DNA*. New York: Atheneum, 1968.

INDEX

Index

Index

Index

Index

Index